DAT
The Complete Guide to Digital Audio Tape

DAT
The Complete Guide to Digital Audio Tape

Delton T. Horn

TAB BOOKS
Blue Ridge Summit, PA

FIRST EDITION
FIRST PRINTING

© 1991 by **TAB Books**.
TAB Books is a division of McGraw-Hill, Inc.

Printed in the United States of America. All rights reserved. The publisher takes no responsibility for the use of any of the materials or methods described in this book, nor for the products thereof.

Library of Congress Cataloging-in-Publication Data

Horn, Delton T.
    DAT, the complete guide to digital audio tape / by Delton T. Horn.
      p.   cm.
    Includes index.
    ISBN 0-8306-7670-8 (hardbound)   ISBN 0-8306-3670-6 (paperback)
    1. Digital audio tape recorders and recording. I. Title.
TK7881.65.H67   1991
621.389'3—dc20   91-9671
                                                                          CIP

TAB Books offers software for sale. For information and a catalog, please contact TAB Software Department, Blue Ridge Summit, PA 17294-0850.

Acquisitions Editor: Roland S. Phelps
Book Editor: Laura Bader
Production: Katherine G. Brown
Book Design: Jaclyn J. Boone
Cover Design: Lori E. Schlosser   EL2
Cover Design and illustration: Greg Schooley, Mars, PA.

Contents

Introduction   *ix*

❖ **1 The History of Sound Recording and Reproduction**   *1*
    Edison's Cylinder Phonograph   *3*
    Improving the Phonograph   *8*
    Spinning Discs   *9*
    Changing Speeds and Times   *12*
    No Mistakes Allowed   *16*
    Disc Materials   *16*
    Meanwhile, the Wire Recorder   *17*
    Tape Recorders   *18*
    Editing Tape   *19*
    Tape to Disc   *21*
    The Shortcomings of Speakers   *22*
    Stereo   *25*
    Reel-to-Reel Recorders   *30*
    The 8-Track Cartridge   *36*
    The Audio Cassette   *38*
    The Quadrophonic Fiasco   *41*
    A Video Diversion   *44*
    Pulse Code Modulation (PCM)   *46*
    The Compact Disc   *47*
    Here Comes DAT   *48*

❖ **2 Analog Sound Recording**   *49*
    The Modern Phonograph   *49*
    Magnetic Tape Recording   *56*
    Wow and Flutter   *74*

❖ 3 Basics of Digital Recording — 75
Numbering Systems *76*
Analog Sound into Digital Bits *80*
Digital Recording Signals *82*
Sampling Frequency *87*
Resolution *90*
Error Correction Codes *92*
Perfect Copies *96*
Multiplexed Signals *98*
Requirements for a
   Digital Audio Tape Recorder *98*
Stationary-Head Recorders *101*
Rotary-Head Recorders *103*
PCM Converters and VCRs *105*
The Advantages and Disadvantages of
   Digital Recording *107*

❖ 4 Compact Discs — 109
Improved Recording *111*
Lasers *113*
Videodiscs and CDs *118*
The Compact Disc *119*
CD Players *123*
Comparing Specifications *128*
The SPARS Code *131*
How a CD is Made *132*

❖ 5 The Basics of DAT — 135
Setting Standards *138*
A Typical DAT Recorder *141*
The Tape *143*
The Head Drum *145*
How Data is Recorded *147*
Signal Sampling *155*
Summary of DAT Specifications *160*

❖ 6 Legal Issues of DAT — 166
The Great Copying Debate *160*
The Notch That Didn't Work *171*
The SCMS Compromise *176*
The Gray Market *178*

❖ 7 DAT Equipment — 180
Automotive DAT Decks *180*
Home DAT Recorders *188*
Digital Audio and Specifications *200*

❖ **8 Maintenance and Troubleshooting** 204
   Care of DAT Cassettes *205*
   Cleaning *210*
   Troubleshooting *212*
   Emergencies *228*

❖ **9 The Future of DAT** 230
   Who's Going to Buy It? *230*
   Continuing Opposition from the RIAA *234*
   The Competition from LPs *236*
   The Competition from Analog Cassettes *238*
   The Competition from CDs *242*
   Conclusions *245*

Appendix   DAT Manufacturers *247*
Index *249*

Introduction

UNTIL FAIRLY RECENTLY, SOUND RECORDING AND REPRODUCTION has been strictly an analog process. Recently, however, the digital revolution has extended into the world of audio. Digital recording offers low distortion, wide dynamic range, flat frequency response over a wide range, and low noise.

In digital recording, analog waves are converted into strings of numerical values which can be recorded, played back, and reconverted back into analog sound waves. Unlike the analog recording techniques of the past, in digital recording, the data can be rerecorded again and again without the slightest bit of degradation in the reproduced sound.

In the 1980s, the digital compact disc (CD) became an exceptionally popular medium for prerecorded music. However, outside of a well-equipped professional recording studio, making your own digital recordings hasn't been possible.

Now, there's a new digital recording format on the scene. It's called DAT, or digital audio tape. With DAT you can make full-featured digital recordings on cassettes that are about one-half the size of standard analog audio cassettes.

After several long years of promises, delays, legal threats, and thwarted hopes, DAT equipment is at long last starting to show up in the U.S. marketplace.

This book aims to be your introduction to this exciting new recording format. What is it? How does it work? What are its special advantages? Why has it been so long in coming to the U.S.? After reading this book, you'll know the answers.

Some early DAT recorders and players from several manufacturers are also discussed in detail in this volume.

Chapter 1 provides a brief history of sound recording since Edison's cylinder phonograph of about a century ago. Chapter 2 reviews the theory and principles behind analog recording. In Chapter 3, the basic principles of digital recording will be introduced, with the compact disc discussed in Chapter 4. The specifics of the DAT format are covered in chapter 5. Chapter 6 looks at some of the legal issues that have held up the introduction of DAT equipment in the U.S. Chapter 7 will examine some of the DAT products that are now or will soon be available.

Chapter 8, on maintenance and troubleshooting, offers many valuable tips on how to keep your DAT equipment up and running at its best performance. Finally, in Chapter 9, we will speculate on the competition faced by DAT, and what the future might hold for this format. An appendix of manufacturers of DAT equipment is also provided.

Digital audio tape is an exciting and revolutionary approach to sound recording. This book is designed to give you the information you'll need to join the DAT revolution.

1
The History of Sound Recording and Reproduction

IN TERMS OF HISTORY, RECORDED SOUND HAS NOT BEEN WITH US very long—only a little over a century. Before that, all music (and speech) was performed live, except for a few automations.

The basic idea of mechanically produced or reproduced music (or other sounds) has held a definite appeal throughout the ages. While in earlier centuries the technology for recording and reproducing sounds did not exist, mechanical contraptions such as music boxes and musical clocks were very popular. Some of these musical automations were quite ingenious and sophisticated in design.

Basically, a musical automation uses some sort of mechanical gearing arrangement or a revolving cylinder with pins to strike vibrating metal tabs or reeds in a specific sequence. Each tab or reed produces an individual note.

In effect, a music box (or other musical automation) might be said to record and reproduce a performance. The music will be performed in exactly the same way each time the music box is used, but the sound itself is not recorded. Each note is reproduced anew for each individual performance. No matter how cleverly designed they may be, music boxes, musical clocks, and other musical automations create each and every sound and note from scratch (according to mechanically preset instructions) each and every time they are heard.

Actual sounds exist only for a very brief period of time. All sounds are nothing but a pattern of vibrations in the air (or some other medium).

A sound is created by forcing something to vibrate. For example, striking a drum causes the taut head to vibrate. In a horn, vibrations are caused by forcing air through a metal or wood column. These physical vibrations are transferred to the surrounding air molecules. When a molecule of air vibrates, it bumps into its neighbor causing that molecule to vibrate too. In that way, the sound is carried outward from the original source.

The vibrating air molecules strike the ear drum of the listener. The patterns of vibration are picked up by the delicate bones in the inner ear. The resulting nerve impulses are decoded by the brain and perceived as sound.

No energy source is infinite, so the vibrations will eventually die away with both time and distance. When molecule A vibrates and bumps into molecule B, molecule B will vibrate with less energy than molecule A. When molecule B collides with molecule C, even less energy will be transferred. Eventually, the molecules will be too weak to cause vibrations in their neighboring molecules.

In addition, no molecule will continue to vibrate indefinitely. When a molecule is forced to start vibrating (by being bumped into or whatever), it has just so much energy. Each vibration cycle uses up some of that energy. Unless new energy is added to the molecule from some external source, eventually it will run out of energy. The vibrations will get weaker and weaker, and eventually they will stop altogether.

Once the vibrations have died out, the sound is gone forever. A similar sound can be created, but the original sound vibrations cannot be stored or retrieved.

Before the technology for recording sound came along, sounds always had very short lifespans. Once a sound had been produced and the last echo had died away, it was gone forever. There was no way to reproduce it.

That changed with the invention of the phonograph in the late nineteenth century. Since then, the craft of sound recording and reproduction has gone through numerous breakthroughs and revolutions.

Currently we are in the midst of the "digital revolution." Digital recording techniques have proven to be a startlingly important breakthrough, permitting new depths of realistic reproduction and low noise and distortion levels. Already, after less than a decade, the digital compact disc (CD) has virtually killed off the old vinyl phonograph album. Despite very vocal resistance by a few diehard vinyl adherents, there is little doubt that the LP is rapidly going the way of the shellac 78 and the Edison wax cylinder. It is well on its way to obsolescence. Times change and no technology can reasonably offer any claims of permanence.

Now, an exciting new development looms on the audio horizon—DAT, or digital audio tape. This is a high-grade, convenient digital recording medium, suitable for use by the average consumer. It promises the best features of both the CD and the audio (analog) cassette.

The big question of the day is, will DAT recorders make analog cassette tape recorders obsolete the way CD players are now making turntables obsolete? Only time will tell. One thing is already perfectly clear, DAT is an exciting new breakthrough in sound recording and reproduction.

Before exploring the specific details of the DAT system, this introductory chapter will quickly guide you through a few of the most important developments in the history of recorded sound. This is not a complete or strictly chronological history. Rather, we will look at a few of the historical paths which have led to DAT.

EDISON'S CYLINDER PHONOGRAPH

The first important sound recording device was the phonograph, developed in Thomas Edison's laboratories in the late nineteenth century. The patent for the phonograph was applied for in late 1877, although work on the project certainly began somewhat earlier than this. The precise origins of the idea and the details of early unsuccessful prototypes are lost in the mists of history.

The system, while crude, was practical, even though many people (including Thomas Edison himself) considered it little more than an amusing novelty. The possibilities of the modern audio industry were inconceivable at the time. The closest thing to recorded music available was a music box, which was considered just a toy or novelty. Nice, but not a major threat to the sheet music business. In those days, when the general public wanted music in their home, they sat down at the family piano, or got some friends together, and played it themselves. There was no alternative. The idea of playing recorded music in the home (let alone, a multimillion dollar recording industry) was not rejected so much as it simply wasn't conceivable. It was too far removed from anything that had existed up to that time for the idea to even occur to anybody.

The possibility of recording music was considered, of course. After all, sound is sound, and once voice recording was possible, it didn't take a great deal of imagination to think of recording music too. But the prevailing opinion of Edison's time was that music recordings would never be anything but a flashy gimmick. Why would anybody really want a recording of music? Live musicians would always be better and more practical. This was the unquestioned reasoning of the day.

Initially, Edison was extremely skeptical of the potential commercial possibilities of the phonograph, and he discouraged his assistants from wasting time on this silly little invention. But when a practical prototype was produced, he figured he might as well patent it, just in case.

Once Edison had a patent on the phonograph, he turned his attention to finding possible commercial applications for the gadget, although he never marketed the phonograph with much enthusiasm. Edison very strongly pooh-poohed the idea of commercially produced musical recordings. He saw the practical applications for the phonograph more in the area of business than in creating a new and powerful entertainment industry.

Some of Edison's competitors had a bit more foresight in this area than he did, and they started to release cylinders of recorded music. When it became clear there was a market for this kind of thing, Edison jumped on the bandwagon too; although, again, never with much enthusiasm. It is clear that the phonograph was never one of Edison's favorite inventions, even though it is one of the best known and most popular of all the devices that came out of Edison's laboratories.

Edison's cylinder phonograph machine was a quite simple mechanical device. In a way, it's so simple, it's a little surprising that no one hit upon the idea earlier. Actually, it's not at all unlikely that one or more early inventors played around with similar ideas, but didn't see the possibilities of practical applications any better than Edison did. Edison was just the first one to bother patenting the device, which probably had more to do with Edison's practice of accumulating as many patents as he possibly could, whether he thought the patented invention was of any real use or not.

The operating principles of the Edison cylinder phonograph are fairly simple. During recording, acoustic sound waves are picked up by a relatively large cone. It had long been known that such a cone could be used to focus and direct sound waves. Similar cones had often been used as very primitive hearing aids (see Fig. 1-1).

At the narrow base of the cone is a stiff needle, or stylus. The sound waves are picked up by the wide opening at the outer end of the cone. As the sound waves pass through the cone, it is forced to vibrate in step with the sound waves. The vibrations are passed along the length of the cone to the stylus at the cone's focal point.

The tip of the stylus is pressed against a cylinder wrapped in tin foil, which is mechanically revolved as it is slowly moved from left to right, rather like the platen of a typewriter. The vibrating stylus etches out a squiggly continuous line in the tin foil coating the cylinder. The groove is etched out vertically. The stronger the vibration, the deeper the groove is cut into the tin foil. This squiggly line was a crude approximation (or analog) of the sound picked up by the cone.

Figure 1-1 *It has long been known that a cone can be used to focus and direct sound.*

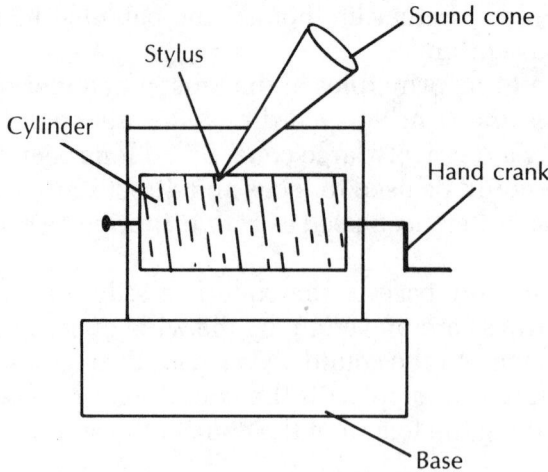

Figure 1-2 *The basic Edison phonograph was simple, but functional.*

The basic Edison phonograph is illustrated in Fig. 1-2. The same set up was used for both recording and playback.

In some machines a different, softer stylus was used for playback, while in other phonographs the same stylus was used for both recording and playback. A loud noise or other vibration of the cone and stylus assembly during playback conceivably could be added to the sound recorded on the cylinder. Both sounds would then be permanently etched into the tin foil wrapped around the cylinder.

For playback of the recorded sound, the cylinder is returned to its starting position, and the stylus is placed at the beginning of the groove etched into the tin foil. As the cylinder is rotated, the stylus vibrates up and down, following the etched groove. This vibration of the stylus is passed through the cone, and therefore the air at the wide end of the cone. The cone now acts like a crude loudspeaker, reproducing the sound recorded on the tin foil cylinder.

People of the day were stunned by the accuracy of the reproduced voice when they listened to Edison's phonograph. (Early cylinders were almost always recordings of the speaking voice. The idea of recording music came a little later.) Some people insisted there was somebody in hiding, speaking the recorded words. Others were sure this contraption was the work of the devil or, at least, spirits of some sort—it couldn't be natural!

If you built a phonograph just like Edison's today, you'd probably consider the reproduced voice to be barely recognizable. The playback was extremely noisy, and at a very low volume. (Electronic amplification was still a few years in the future.)

The universal lack of awareness of the potential market for music recordings wasn't really so bad, because commercial quality musical recordings were not yet very practical. The available technology was not yet up to the task, although it would catch up very quickly.

Another significant limitation of the early phonographs was that they were powered by a manually turned crank. Winding the crank too fast or too slow during either recording or playback would change the pitch and speed of the reproduced sounds. Some later more sophisticated cylinder phonographs used a wound spring mechanism (rather like a watch) for a somewhat more consistent running speed. A wound spring mechanism was a definite improvement over a hand crank, but considerable speed fluctuations were still the norm. If the spring was wound too tightly, the cylinder would be rotated too rapidly, resulting in a chipmunk-like high pitch. On the other hand, as the spring mechanism began to wind down, the speed would drop, turning the reproduced sound into a slow, low-pitched groan. Both the hand crank and the spring motor were carried over into the early turntables, which will be introduced later in this chapter.

By modern standards, the Edison cylinder phonograph reproduced sound very, very poorly. At the time, however, many people were extremely impressed with the incredible accuracy of the gadget's sound reproducing capabilities. The whole concept of recorded sound was so new, they really had nothing to compare the phonograph to.

It is rather amusing to consider how the human ear seems to grow more sophisticated with each new development in sound recording technology. Each new breakthrough has been hailed as the most realistic reproduction yet. Enthusiasts swear the recorded results are virtually indistinguishable from the original live sound. Listeners from a later generation are always amazed at the enthusiasm of their parents over such pathetically low fidelity sound. Will our grandchildren be equally amused by how impressed we were with current digital recording techniques?

IMPROVING THE PHONOGRAPH

Almost as soon as the phonograph was invented, a number of ingenious people went right to work improving the system. It's hard to imagine that such a crude device as the phonograph could have been the beginning of a direct line to the new, highly sophisticated DAT machines, but it's true.

One of the first improvements was to replace the tin foil coating of the cylinder with wax. This permitted finer, closer spaced grooves, increasing the recording time and fidelity somewhat, although the reproduced volume was lower with a wax cylinder than with a tin foil cylinder.

As the potential profits promised by the phonograph became obvious, there were many experimenters looking for better materials to use for the stylus and the cone. The size and shape of the cone was also experimented with.

Probably the most important developments lay in the mechanism for rotating the cylinder. More precise gearing allowed adjacent grooves to be etched side by side, increasing the recording time for a cylinder of a given size. The earliest phonograph cylinders held just a few seconds of sound. One of Edison's earliest test cylinders was just long enough for him to recite the first verse of "Mary had a Little Lamb."

Eventually, the phonograph was improved enough that cylinders as long as 2 or 3 minutes were feasible. Somewhere along the line, somebody got the bright idea that this was long enough to record a song. Many important vaudeville stars and a few opera stars recorded phonograph cylinders. There was no way to mass produce or duplicate cylinders. The performers had to repeat the song over and over to produce multiple copies.

Some of these early phonograph cylinders have been carefully preserved and still exist. These very early recordings have been converted to more modern systems, including CDs. Curiously, with modern noise reduction systems, we may be able to hear clearer reproduction of these recordings than their original owners.

The audio industry had been born. Still, it probably would not have grown much beyond a novelty or a passing fad unless a major breakthrough got rid of the cylinder and replaced it with a platter, or disc.

SPINNING DISCS

The wax (or foil) cylinder suffered many inherent limitations, including poor fidelity, short recording times, uneven speeds, awkward storage, and inconvenient duplication of recordings. Many (though not all) of these problems could be reduced, or almost eliminated, by flattening the cylinder into a circular disc, as illustrated in Fig. 1-3. This development occurred in the 1890s, the invention of a man named Emile Berliner.

Besides flattening the recording medium, Berliner also changed the way the grooves were cut. The early Edison cylinders used a vertical, up and down groove. Berliner's discs used a vertical back and forth (laterally zigzagging) groove.

To limit confusion with the Edison phonograph, Berliner called his device a gramophone. The word gramophone looks a little odd to most modern Americans, but it's really pretty straightforward. *Gram* means write, like in telegram, and *phone* means sound. So a gramophone is just a device for writing (recording) sound. Similarly, the term Edison used—phonograph—means exactly the same thing. *Phono* is a variant on phone (sound) and *graph* also means write. So we replaced the sound-write (phonograph) with the write-sound (gramophone). In England (and certain other areas), the newer term took hold and is still used, while in America, the older term for cylinder machines ended up being applied to disc machines too. The English say gramophone, while Americans say phonograph. In modern times, the two words are pretty much interchangeable.

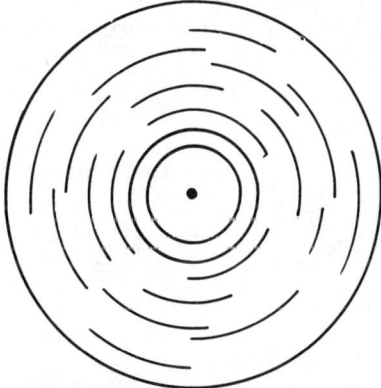

Figure 1-3 *Berliner flattened the Edison cylinder into a more easily reproducible disc.*

In the disc recording system, we still have a continuous vibrating groove cut into the platter's surface by a stylus. Instead of loops around a three-dimensional cylinder, the groove now takes the form of a continuous spiral, starting with a large circle at the outer edge of the disc, and moving inward to a smaller circle near the center.

A small hole is drilled in the center of each disc. This hole is fitted over a spindle to hold the disc in place on a revolving platform, or turntable. The stylus, or needle, is mounted on the end of a pivoted arm, called the tonearm. At the beginning of the record, the stylus is placed in the outermost groove. As the disc revolves, centrifugal force eases the stylus inward towards the center, following the squiggly spiral etched into the platter's surface.

In the earliest (nonelectronic) devices of this type, a sound cone (just like the ones used in the cylinder machines) was mounted on the tonearm.

The basic features of an early disc playing machine are illustrated in Fig. 1-4. At first, similar equipment was used for recording and playing back discs, just as was the case with the cylinder machines. As the recording industry started and became more sophisticated, subtle but increasingly important differences between recording and playback machines started to emerge. The phonograph quickly became a playback-only medium for consumers. (Although a few home recorders for discs were marketed as late as the 1950s.)

Being a flat disc, rather than a three-dimensional cylinder, the new type of recordings were much more space efficient. A longer recording groove (and therefore a longer playback time)

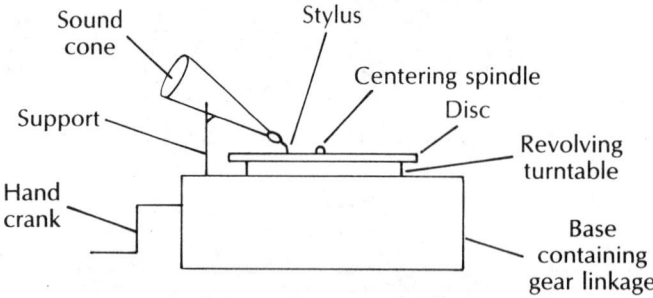

Figure 1-4 A Berliner gramophone wasn't much more complex than an Edison phonograph.

could fit onto a disc that was easier and more convenient to store. In addition, recording on both surfaces (sides) of a disc was possible, doubling the available recording time overall.

The flat discs also proved to be much more convenient to reproduce commercially. The song (or whatever) only had to be recorded once, and then copies were made of the original disc. Basically, a mold was made of the original (master) disc, and then numerous duplicates could be inexpensively pressed out from the mold. Of course, as the recording industry grew, the process became more complex and sophisticated, but the principles remained the same, right up to today.

The earliest disc phonographs had to be manually cranked during the entire record. If you stopped cranking, the disc stopped spinning. Needless to say, recording and playback speeds varied wildly. Many early disc recordings played back on modern (and precise) turntables run too fast or too slow. This same thing shows up in many of the old silent movies. The camera was hand cranked, and human beings are capable of just so much accuracy in the speed of winding a crank, so when projected by an automatic, steady-speed projector, the action appears jerky and irregular.

Spring motors (similar to watches) were soon added to phonographs, which was a major convenience. Now, you just had to wind up the machine before you played the record. You could sit back and listen, and give your arm a rest. Of course, if you didn't do enough winding, the motor would run down in the middle of the record, slowing to a disharmonious stop until you wound the crank some more, which usually restarted the playback at chipmunk-like speed before the mechanism stabilized. Even at its best, this type of motor never offered much accuracy in the turntable speed. Still, it was better than continuous manual cranking, and there really wasn't anything better available for a number of years. Electricity was not yet in widespread use, especially in the average home. The recording industry had to wait until the 1920s before electrical recording was a practical option.

Eventually, of course, electrical power became the norm, and phonographs used electric motors, which required a lot less work—you just had to flip a switch—and were considerably more accurate. The earliest models were still very inaccurate by modern standards. A turntable designed for 78 rpm (revolutions per minute) might actually spin at anywhere from about 65 to 90 rpm, but even this was a big improvement. Besides, the same speed irregularities showed up in the recording machines. If you were lucky, your favorite record was recorded on a machine with the same speed as your phonograph.

CHANGING SPEEDS AND TIMES

At first, phonograph discs were recorded at almost any speed. It depended on how fast the operator turned the phonograph's hand crank while the recording was being made. During playback, the operator had to adjust his cranking speed until the reproduced voice or music sounded right. This was obviously quite inconvenient and awkward.

Things were a little better with phonograph machines using a spring motor, but there was still no standardization. If the motor spring in your playback machine didn't match that of the recording machine, the disc would be played back too fast or too slow.

Even when electrical motors were first used to drive turntables, speed standardization was still uncommon. Many early electrical turntables featured a speed control rheostat so the playback speed could be more or less matched to the recording speed. For a long time, there was simply no standardization in disc speed at all. This situation was adequate for experimentation and occasional novelties, but for a real consumer industry to grow, some sort of standardization was essential.

Disc speed is defined by how many times the turntable turns completely around in 1 minute. This is called rpm, or revolutions per minute. The faster the platter revolves, the better the fidelity. This will be explained in more detail in Chapter 2, but basically, the faster the speed, the longer the groove segment for a given note. With a longer groove, there is obviously more room to record greater detail.

Many different standard speeds were proposed and tried in the early years of the budding recording industry. The one that eventually caught on as the nominal universal standard was 78 rpm. At first, as noted earlier, this was really just a nominal standard. Great precision in the disc speed was not expected, nor, considering the technological level of the available equipment, was such precision very practical or even possible. Anything in the ballpark was considered acceptable.

Once at least rough standards were put into place, the disc recordings proved to be remarkably popular. Literally millions of 78-rpm recordings were sold over the years. Many 78-rpm discs still survive today. Most are now being protectively hoarded by collectors, but there are undoubtably still a great many quietly mouldering away in forgotten boxes in hundreds of attics and basements.

Disc sizes also varied a great deal, but 12 inches (diameter) soon became the unofficial industry standard. Seven-inch discs were also common for shorter recordings. A typical 12-inch disc averaged about 7 minutes of music (or whatever) on each side. This was more than enough for most popular songs. In fact, a smaller 7-inch disc could usually hold one complete popular song per side. A 12-inch disc could hold two or three short songs on a single side.

This was fine for pop selections, but serious classical music—symphonies, concertos, etc.—presented something of a problem to producers of 78-rpm records. Longer pieces just couldn't be squeezed onto one side of a normal disc. Such major works had to be broken up over two or more disc sides. Multiple disc sets were called albums, and were usually released in book-like folders with a separate bound-in sleeve for each disc of the set. A typical album would contain a full-length symphony or concerto, or a collection of related shorter pieces. Obviously, complete opera recordings were a real rarity at this time.

To minimize the distraction and inconvenience of having to change the record every 7 minutes in a long piece of music, some turntables featured automated record changers. Two or more discs were stacked up on a high spindle. When one disc was finished playing, the tone arm automatically moved out of the way, and the next disc dropped down (often quite roughly) onto the turntable. The tone arm moved back into place, and the music resumed. This development showed up primarily in electrically powered phonographs. It just wasn't practical with the earlier mechanisms.

To accommodate automatic changers, record sides in multi-disc albums were often numbered in a seemingly odd fashion. For example, in a four disc set, the flip side of side 1 would be side 8. Side 2 would be paired up with side 7, and side 3 with side 6. Finally, sides 4 and 5 shared a disc. The discs were stacked up on the changer to play sides 1 through 4 in order, then the pile was turned over, permitting sides 5 through 8 to be played in the proper order.

Even while it became the standard, 78 rpm certainly wasn't the only speed used. When less fidelity was required, such as spoken word recordings, a lower speed could be used. For spoken word recordings 16 rpm became moderately popular. A typical 16-rpm disc side could play about 30 minutes. The sound was pretty lousy, even by the standards of the day, but it would do well enough for simple and noncritical spoken word recordings. Many "talking books" (a voice reading a book out loud) were released on 16-rpm discs.

As the electronic era set in, fidelity was improved considerably. The sound could be electronically amplified, and tone controls were added to both the recording and the sound reproduction systems. These developments will be discussed in more detail in Chapter 2.

With the improvements offered by electronic amplifiers, better sound quality could be obtained with slower disc speeds. Two new standard speeds appeared, almost simultaneously.

The 45-rpm disc, championed primarily by RCA, was usually 7 inches in diameter, and was designed to hold a single song on one side. A 7-inch 45-rpm disc held about the same amount of music as a 12-inch 78-rpm disc.

To avoid confusion, 45-rpm discs featured a considerably larger center hole. Curiously, this mechanical differentiation was never employed with other standard disc speeds. In fact, it proved to be something of a nuisance. A special wider spindle was needed in order to play 45-rpm discs. Once the format caught on, most turntables came with a secondary spindle that slipped over the normal smaller spindle. This was an inelegant solution, but it was inexpensive. Many record player owners faced considerable frustration whenever they mislaid the 45-rpm spindle adapter.

A number of manufacturers and dealers soon offered a slightly more efficient solution to the incompatible spindle size of 45-rpm discs. Small plastic (or occasionally metal) discs could be inserted snugly into the large hole of a 45-rpm disc, leaving a smaller hole suited for use with a standard turntable spindle.

It would certainly seem that RCA made a mistake by making their 45-rpm discs mechanically incompatible with existing discs. On the other hand, the large hole 45-rpm discs seemed to have a certain appeal all their own, especially for teenagers—the primary purchasers of this format. Somehow, these 7-inch discs with the large hole seemed "friendlier" than standard 12-inch albums with their smaller spindle holes. This appeal might make for a very interesting psychological study.

About the same time the first 45-rpm discs were commercially released, the 33⅓-rpm disc was introduced by Columbia. It was usually 12 inches in diameter, like a standard 78-rpm disc, but the slower speed allowed much longer recording times—up to about 20 minutes per side. (Later refinements permitted even longer times on the same sized disc.) Because of the greatly increased recording time, the 33⅓-rpm discs were dubbed LPs (long playing discs). LPs were first marketed in the 1950s, and rapidly grew in popularity enough to truly dominate the recording industry for about one-quarter of a century.

The new LPs were mechanically identical to the earlier 78-rpm discs. The same disc size was used (12 inches), and the spindle hole was exactly the same size. No mechanical spindle adapter was required.

Several songs, or even a moderately long classical piece could fit onto a single disc. Because one 33⅓-rpm LP could hold the equivalent of a multidisc 78-rpm album, single discs of this slower speed came to be called albums too.

The 33⅓-rpm disc's longer playing time led it to rapidly become the most popular disc format and the de facto standard. The 33⅓-rpm LP quickly grabbed the lion's share of record sales.

On the other hand, 45-rpm discs were very convenient for use on popular music radio stations, and were sold as "singles," mostly to teenagers who were only interested in the hits they heard on the radio. The 45-rpm disc was also quite well suited to the sock hops so popular in the late 1950s and early 1960s.

Meanwhile, as the 33⅓-rpm LP and the 45-rpm single gained popularity, the 78-rpm disc rapidly faded into obsolescence. It couldn't live up to the competition of the new, improved disc formats.

The 33⅓-rpm disc "ruled the roost" in consumer music reproduction through the 1950s and 1960s. It's predominance started to be eroded somewhat by the audio cassette in the 1970s, although the vinyl LP continued to be the most popular commercial recording format. Then in the 1980s, this long popular format lost its dominant position in the marketplace, and was virtually killed off by the CD. The 33⅓-rpm disc is not totally dead yet, but virtually all observers of the audio industry agree it is rapidly following the 78-rpm disc into the mists of history.

NO MISTAKES ALLOWED

In a way, during the early days of the recording industry, the fairly short times possible on 78-rpm discs were something of an advantage. No editing at all was possible. Everything recorded on the disc was performed live from start to finish. It was not even possible to record part of the disc, stop, and later restart the recording process. Once the process of cutting a disc was begun, you had to see it through to the end.

If any of the performers made a mistake, or if there was a loud, extraneous noise, there were just two choices available to the early record makers: leave the disruption in the final record, or throw out the disc and start all over from the beginning with a new platter.

Fortunately, with only about 7 minutes of recording time on a side, this usually wasn't too serious of a problem. But as recording times got longer, the lack of editing capabilities became increasingly undesirable.

DISC MATERIALS

The earliest recording cylinders used tin foil as the recording surface. Later cylinders were coated with wax. When the recording industry moved on to flat discs, manufacturers experimented with several materials. Through most of the 78-rpm disc's reign, the standard platter material was shellac. Shellac discs were thick, heavy, and quite fragile, but they provided quieter and more reliable recording surfaces than wax or foil. The discs were also far easier to store, ship, and reproduce than the cylinder.

About the time the LP (33⅓-rpm disc) was developed, plastic technology was becoming popular and these new "wonder materials" were being used in countless applications by many different industries. Shellac platters were quickly replaced with vinyl discs. Vinyl discs can be made thinner and more flexible (and therefore less prone to breakage) than the old shellac discs. More importantly, a vinyl surface can be made much smoother than a shellac surface. This means, the new vinyl discs were far quieter, with considerably less in the way of surface noise—the clicks, pops, and constant "swishing" that plagued most shellac 78-rpm recordings.

Certainly the new vinyl LPs still had their fair share of clicks, pops, and swishing, but in comparison with their noisier predecessors, the newer discs were a very marked improvement. Many enthusiasts were convinced that the quality of recorded sound had reached its ultimate level of perfection. Such optimism frequently shows up throughout the history of recorded sound. Later developments always make the previous enthusiasts look foolish and naive. The current enthusiasts of state-of-the-art digital equipment should keep this fact of history in mind.

Over the relatively long lifespan of the vinyl LP, a number of improvements were made in the basic format. Higher-grade vinyl and new, improved variants on the basic chemical offered improved sound quality (less surface noise) and better disc durability.

Some record manufacturers, catering to the serious audiophile market, advertised that their discs were pressed only from virgin vinyl. Recycled vinyl was likely to contain impurities that could adversely affect the smoothness of the disc's surface. The less smooth a disc's surface is, the greater the amount of inherent surface noise it will have.

MEANWHILE, THE WIRE RECORDER

The history of sound reproduction equipment is rather complicated. It does not follow a single, straightforward line of technological development.

So far we have concentrated on phonograph and disc recordings, but now we'll take a detour back to the 1940s, when a completely different approach to recording sound was developed. This new system was magnetic recording, and it works on entirely different principles than phonographs.

As unlikely as it sounds, the earliest magnetic recorders

used a simple spool of wire as the recording medium. As was the case with the phonograph, initial fidelity for magnetic recordings was rather poor, and applications were limited primarily to voice recording. Actually, wire recorders never enjoyed a significant improvement in sound fidelity, but they led directly to tape recorders, which did grow to offer excellent high fidelity reproduction.

During the war, wire recorders were used primarily as a propaganda tool in Nazi Germany. The idea was to confuse the enemy about the whereabouts of political and military leaders. A plot to assassinate Hitler, for example, wouldn't have much chance for success if nobody could figure out which city Hitler was in. A recording of an important leader would be played on a radio station in area A to conceal the fact that the speaker was really in area B. This was a very simple but effective trick, since most of the world did not know that the necessary recording equipment existed.

In a gramophone or phonograph, an analog representation of the recorded sound exists as a groove physically cut into the recording surface. Anyone could look at a disc and easily tell if it had anything recorded on it or not. There is no visible change in magnetic wire as it is recorded on. The signal is electrical rather than mechanical in nature. An analog representation of the recorded sound is stored as a pattern of varying magnetic charges along the length of the wire. We'll go into a little more detail on the principles of magnetic sound recording in Chapter 2.

TAPE RECORDERS

Wire recorders were functional but severely limited in the potential fidelity of their sound reproduction. Improvements were soon made to the basic concept of magnetic recording. The primitive wire recorder was useful for the Nazi's propaganda tricks, but not much else.

A wider recording area (permitting more space for detail) was provided by replacing the wire with a strip of tape. Wire recorders were soon replaced by tape recorders, which operated on the same basic principles, but with much better results. The relationship between the wire recorder and the tape recorder is remarkably similar to the relationship between the cylinder phonograph and the disc phonograph.

As practical tape recorders were developed, several widths of tape were used, but one-quarter inch became the standard. One-half-inch and 1-inch wide tapes are often used in professional recording applications, where more tracks and higher recording detail are required.

The recording tape consists of a strip of nonactive base material coated with a magnetically sensitive oxide material. The coating, which is the actual recording medium, contains many tiny magnetic particles suspended in a special gluelike substance known as the *binder*.

Early tapes used paper as the backing base material. This paper-based tape was eventually replaced by cellophane, then various plastics, such as mylar. Improvements were also made in the magnetic oxide coating.

Like early phonograph records, early magnetic tapes generated quite a bit of built-in noise and distortion. Improved tape formulations have reduced such problems considerably.

As with a phonograph disc, there is a trade-off between fidelity and recording time depending on the speed used. Tape speed is measured in how much tape passes over the recorder's head (or heads) in a second. The speed figure is given as IPS, or inches per second.

The faster the tape speed, the greater the recorded detail; thus, the better the fidelity of the reproduced sound during playback. On the other hand, slower tape speeds permit longer recordings on a given length of tape.

Standard tape speeds are related by factors of two. The next higher speed is twice as fast as the next lower speed. The most common tape speeds in use today are 1⅞ IPS, 3¾ IPS, 7½ IPS, 15 IPS, and 30 IPS.

Of these 3¾ IPS and 7½ IPS became the most widely used speeds for consumer (home) tape recording equipment, with 7½ IPS as the nominal standard. In professional tape recording studios, speeds of 15 IPS and 30 IPS are generally used, with 15 IPS as the usual standard.

EDITING TAPE

Tape recorders offer a couple of important advantages over phonograph records. For one thing, a magnetically recorded tape can be erased, then reused to record completely different sounds. Tape can also be easily edited.

There are two basic types of tape editing. One method is to erase a portion of an old recording and record new material over the top of it. There are several possible variations on this idea. Several of these will be discussed briefly in Chapter 2.

More commonly, tape editing involves physically cutting and splicing the tape, as shown in Fig. 1-5. Say there are three sounds recorded on a strip of tape. We'll call these three sounds A, B, and C. In the figure, we cut the tape between A and B, and between B and C. We then discard the piece of tape containing sound B and join (splice) tapes A and C together. The tape now contains sound A, followed by sound C. Sound B is gone and will not be heard when the edited tape is played.

The same trick can be expanded slightly to alter the order of sounds recorded on the tape. In this case, let's assume there are four separate sounds recorded in sequence, which we'll call A, B, C, and D. Now, we'll make the following cuts in the tape; between A and B, between B and C, and between C and D.

Now, we carefully reverse the positions of the tape segments containing sound B and sound C. When we splice the pieces of tape and play it back, we find the sounds now appear in this order: A, C, B, and D.

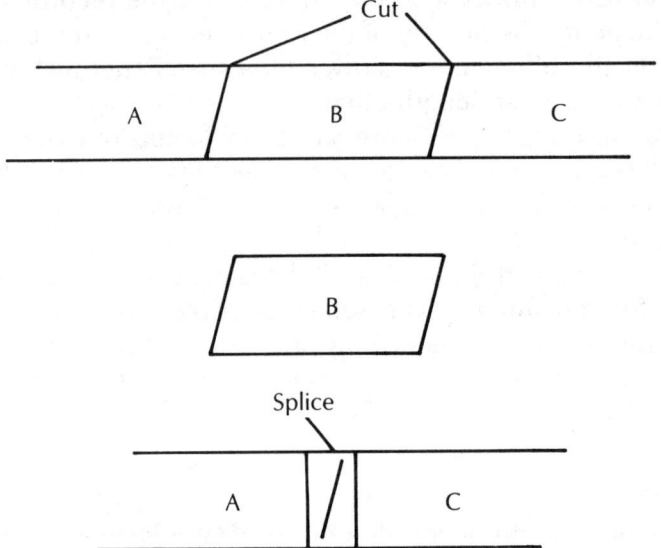

Figure 1-5 *Analog tape can easily be edited by making physical cuts and splices.*

Obviously, this type of editing permits much greater flexibility in the recording process. With tape recorders there is no need to record entire pieces straight through from start to finish, without interruption. The piece can be broken up into convenient segments and reassembled in editing. Different parts can be recorded separately, and the tape can be edited into a complete performance that never actually took place.

In editing tape, the cut is made at an angle to avoid creating a loud "pop" on the tape as the spliced joint passes over the playback head. The splice is made by abutting the two angled edges of the tape segments to be spliced together and covering the joint with a special splicing tape. Do not ever use ordinary cellophane tape to splice a recording tape. The adhesive on the cellophane tape can bleed out under the edges, causing adjacent layers of tape on a reel to stick together. It can also gunk up the heads and tape path of the tape recorder. It's not worth cutting corners here. Use tape designed expressly for the purpose at hand.

TAPE TO DISC

The editing capabilities of tape recording are so powerful and useful that most commercial recordings are first done on tape, and are then transferred to disc (after editing) for commercial release. Virtually all LPs were made in this manner.

A rather curious exception to this approach appeared in the late 1970s. Several recordings marketed to fussy audiophiles were recorded "direct to disc." This was done to minimize noise and hiss from analog tape. Also, purists disliked the idea of an edited performance. It was a great irony that one of the new technological advances in audio was really a reversion back to the primitive practices of the 78-rpm days.

Direct-to-disc recordings never really took off in the marketplace. They were more expensive (and usually shorter) than conventional LPs, and except for some dedicated audiophiles, most consumers really didn't think the improved sound quality justified the increased price. After all, a direct-to-disc LP got scratches, tics, and pops just as easily as any other vinyl disc. The technique was never really very practical.

Besides, before direct-to-disc recordings had time to get any serious foothold in the marketplace, digital recording came along and neatly stole the show. But we won't get into a discussion of digital recording until Chapter 3. The three most popular magnetic tape recording formats will be discussed later in this chapter.

THE SHORTCOMINGS OF SPEAKERS

It is a truism that no chain can ever be stronger than its weakest link. If it's going to break, the weak link is the one most likely to snap.

The same basic idea is true of audio systems. The reproduced sound can never surpass the limitations of the weakest element in the system. Historically, the weakest link in the audio chain has almost always been the loudspeaker, which converts electrical impulses into mechanical sound waves.

A speaker mechanically vibrates in step with the recorded signal causing wave patterns in the air. These wave patterns are interpreted by the ear as sound.

The paper cone in the original gramophones served as the speaker. It was crude and had a severely limited frequency response, but then so did the wax and tin foil cylinders, so the inherent acoustic limitations of the cone didn't really matter very much. That was a good thing because the technology for better speakers did not yet exist.

Generally, the term *speaker* is used for the sound reproducing element in an electrically powered system. The most common type of speaker is the dynamic speaker. In a dynamic speaker, the electrical signal representing the sound to be reproduced is fed through a coil wound near a permanent magnet. When current passes through the coil, a magnetic field is set up. Like magnetic fields repel, and unlike magnetic fields attract. As the current changes direction, the magnet moves closer to or farther from the coil. The magnet is attached to the center, focal point of a paper (or sometimes plastic) cone in a metal frame. The cone is forced to vibrate by the motion of the attached magnet; thus creating sound waves in the surrounding air. (This explanation is very much simplified. We don't need to get more exact about the theoretical operation here.)

To reproduce low-frequency signals, the speaker cone must move fairly slowly, but it must move relatively large amounts of air. The human ear is not a linear device. It does not hear all frequencies equally well. For a given amplitude, or amount of air movement, a high-frequency sound will generally be perceived as louder than an equivalent low-frequency sound.

On the other hand, to reproduce a high-frequency signal, less air movement will result in the same perceived volume, but the speaker cone must vibrate at a much higher rate.

In early electrical phonographs (and even today in some very inexpensive systems), a single speaker is expected to do the entire job. It isn't really very reasonable to expect a single speaker to reproduce all audible frequencies equally well—especially when it has to simultaneously reproduce multiple, widely separated frequencies.

Most modern speaker systems use two or more speaker units in a single housing. Each speaker unit within the system specializes in a specific range of frequencies.

A *tweeter* is a small speaker unit designed to work at high frequencies (treble). The small size permits it to vibrate at high rates fairly easily. Since high-frequency sounds carry fairly well, the tweeter does not have to move large masses of air, so it can be fairly small. Sometimes horn speakers are used in place of the more common cone-type speakers. A horn speaker has greater directionality than a cone speaker. Directionality is far more important for high-frequency signals than for signals at lower frequencies. The human ear is not very good at identifying the location of a low-frequency sound source, but it is very sensitive to the position of a high-frequency sound surce. Horn-type speakers are rarely used for anything but the uppermost frequencies of the audible spectrum.

At the other extreme is the *woofer*. The woofer is normally fairly large. All other factors being equal, the larger a woofer is, the better. It is designed to reproduce low frequencies (bass). It doesn't have to vibrate very fast, so it's relatively large size isn't a problem. On the other hand, bass (low) frequencies generally don't carry very well. More air must be moved to achieve the same perceived volume as a comparable high-frequency signal. So a woofer has to move fairly large amounts of air, and it needs to be pretty large to have enough "oomph."

Better quality speaker systems will often have more than just these two speaker elements (tweeter and woofer). If there are more than two speaker units in the system, the extra units are called *midrange speakers*. A midrange speaker, obviously enough, is between the tweeter and the woofer in terms of size and the frequency range covered.

At the speaker system's inputs, the signal is broken up into the appropriate frequency ranges by specialized filter systems, known as *crossover networks*.

Most modern turntables, receivers, tape decks, and amplifiers typically have very impressive specs (specifications). Generally the distortion introduced by any one device is well under 1%. Tests have indicated that most people can't hear less than 1% distortion. Of course, the distortion introduced by each of the elements in the system is additive. That is, you must take into account the distortion produced by the turntable, the distortion produced by the amplifier, the distortion produced by the speakers, and the distortion produced by the disc itself. This includes all distortion due to any surface irregularities in the disc, along with the distortion produced by the original recording equipment and microphones used to record the disc. Even though a given component, such as an amplifier, may have an inaudible level of distortion (1% or less), it may have an audible effect on the final reproduced sound once it is added to the distortion from all of the other links throughout the entire audio (recording and reproducing) chain.

Even good, expensive speakers, typically introduce as much as 3% distortion. As you can see, the speaker is definitely the weak link in the audio reproduction chain.

It is extremely difficult to define meaningful specifications for a speaker system. This is because the speaker inevitably interacts with the room it is used in. A speaker system that sounds great in room A may sound lousy in room B. This is because the physical dimensions and hard (reflective) and soft (absorbtive) surfaces in the room emphasize certain frequencies, while decreasing signals at other frequencies. The ideal is perfectly flat response; that is, all audible frequencies are equally reproduced without peaks (overemphasis) or valleys (de-emphasis).

Because the listening room has such an impact on the frequency response of the speaker system, how can general frequency response measurements be made? Often, the frequency response is made in anechoic (echo-free) chamber. Because the chamber is acoustically dead, it does not affect the speaker's operation. Unfortunately, it is highly debatable that the results of any frequency response measurements made in an echoless chamber have any meaningful bearing on real-world results. If a speaker had a perfectly flat response in the echoless chamber, it's frequency response would inevitably be decidedly nonflat in any practical listening environment.

So far, no one has found a truly practical way to define the frequency response of a speaker system. In shopping for speakers, you must rely on your own ears. It would be quite safe and reasonable to simply ignore any frequency response graphs provided by the manufacturer. You can practically guarantee such a graph will be completely meaningless in determining how the speaker system will actually sound in your living room. If possible, listen to the speakers in your intended listening room before buying them. This isn't always possible, of course. The next best thing is to make sure that the dealer has a reasonable return policy. If you get the speakers home and decide they don't sound good in your listening room, the dealer should refund your money, or allow you to exchange them for a different set of speakers.

To further complicate matters, no human being's ears have a flat frequency response. I've seen many audiophiles get into long, drawn-out arguments over which of two speaker systems sounds better. Joe likes speaker A much better than speaker B, while George insists speaker B sounds a lot better than speaker A; and both Joe and George are absolutely correct. The difference of opinion stems from minor (but critical) differences in the physical mechanisms of their own ears.

STEREO

There have been several important breakthroughs in the field of audio recording, each altering the face of the industry. Audio recording started with the cylinder gramophone in the 1880s and became a practical industry with the 1890s. Great strides forward were taken with the advent of electrical recording in the 1920s and magnetic recording in the 1940s. Two important and revolutionary developments showed up in the 1950s. One, the LP, was discussed earlier in this chapter. The other was perhaps the most revolutionary of all the developments in the entire history of audio recording—stereo.

From the first wax cylinders recorded in Edison's laboratories, all recorded sound was monaural, or monophonic. These

26 The History of Sound Recording and Reproduction

terms are essentially interchangable. In both, the prefix *mono-* (or *mon-*) means one. The term *-aural* means of the ear, or having to do with sound. Similarly, *-phonic* means having to do with sound. Thus, both *monaural* and *monophonic* mean one sound source. Both terms are normally abbreviated as just mono.

A mono sound system has just one speaker (or speaker system)—separate woofers and tweeters don't count as individual speakers in this sense. In effect, listening to a monophonic system is a little like listening to music from the next room through a hole in the wall. All sounds come from a single point in space, as illustrated in Fig. 1-6.

With the development of the LP and improvements in electronic circuitry, such as amplifiers, more realistic sound reproduction became possible. Good quality sound systems began to be labeled high fidelity, or hi-fi for short. With the birth of hi-fi, serious audiophiles began to appear. An *audiophile* is a person who seeks the "holy grail" of perfect sound reproduction. Many truly dedicated audiophiles would rather go hungry than go without the latest state-of-the-art audio equipment.

Hi-fi monophonic sound reproduction got quite good, considering the limits of the technology. But even if separate woof-

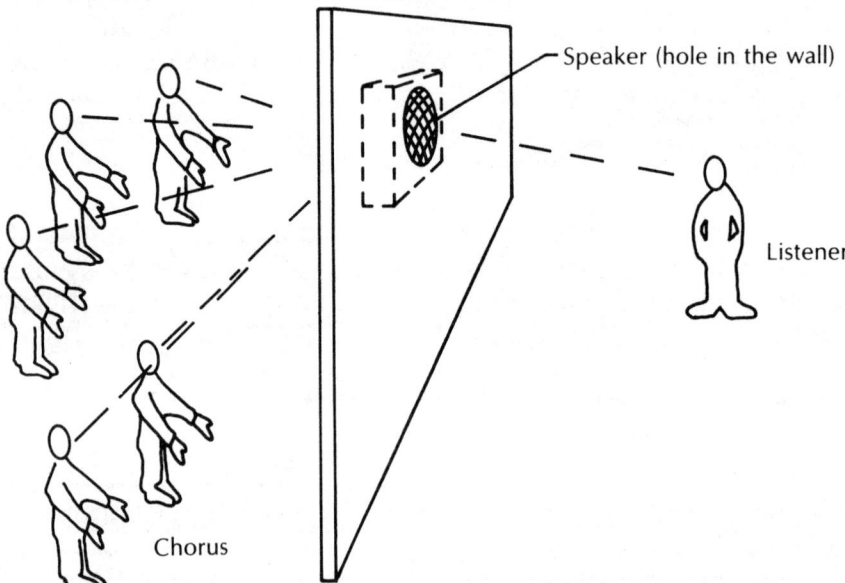

Figure 1-6 *A monaural sound system gives the "hole-in-the-wall" effect.*

ers and tweeters were used, they were mounted in a single cabinet, and were effectively the equivalent of just one speaker. So all of the sound, no matter how good it might be, came from a single point in space—the hole-in-the-wall effect.

Real world (unrecorded) sounds, however, arrive at our ears from varying directions. With two ears, we can usually pinpoint the source of a sound with reasonable accuracy. Thus, a monophonic hole-in-the-wall recording tends to be a bit lacking in realism, even though the sound quality in terms of frequency response and distortion may be extremely good.

In the 1950s someone got the bright idea that since we have two ears, why not use two speaker systems, each producing half the total sound. Some sounds would come from the right speaker and others would come from the left speaker. This was certainly a logical sounding idea. The result is called a stereophonic, or simply a stereo sound system.

At first glance, it might seem that a stereo system wouldn't be that much of an improvement over a good quality monaural hi-fi system. It would appear to be nothing more than two holes in the wall instead of one, and there would be nothing in between the two holes. Fortunately that isn't the way things work out in practice. You only get the two isolated holes-in-the-wall effect if there is no duplication between the signals fed to the two speakers. This rarely occurs in practice, except for intentional special effects.

By splitting up the signal between the two speakers, a sound can be located anywhere between them. A stereo system can apparently place the recorded instruments or voices in a number of different positions, as illustrated in Fig. 1-7.

To understand how the stereophonic effect works, let's take a look at a simple example. Let's suppose we are making a recording of a trio consisting of a violin, a cello, and a viola. We will assume that each of these three instruments plays a brief solo passage, each at the same volume, so that if this was a monaural recording the speaker would be fed a constant power level, say, 3 watts.

For our stereo recording, we will record the violin so that its full 3 watts is applied solely to the right speaker, with none of the signal going to the left speaker. The violin's apparent acoustic location is to the far right of the listener.

Similarly, for the cello's solo, the left speaker will receive

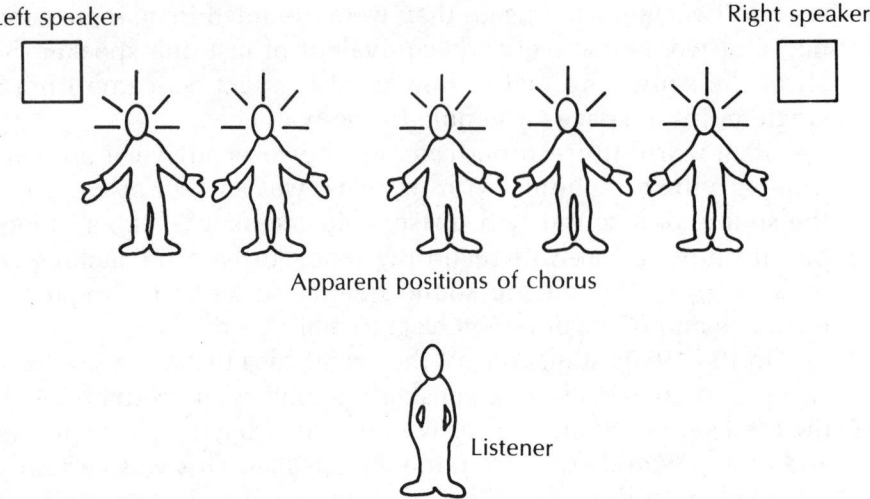

Figure 1-7 A stereophonic sound system gives an apparent spread-out sound stage.

the full 3 watts from the amplifier, and the right speaker will receive no signal. In this case, the cello's apparent acoustic location is to the far left of the listener.

Well, so far, the effects of adding the second speaker to the audio reproduction system is certainly no big deal. We just have the predicted two holes-in-the-wall effect. Not much of an improvement there.

The viola solo, however, is recorded so that each of the speakers receives half the signal. The right speaker receives 1.5 watts and the left speaker receives 1.5 watts, for a combined total of 3 watts. On playback, both speakers will be reproducing the same signal. Since the sound coming from the two actual sources (speakers) is identical, our ears will interpret this sound as coming from a nonexistent single source at the midpoint between the two speakers. The acoustic image would appear to resemble the drawing in Fig. 1-8.

In this manner, various sounds can have apparent locations anywhere in the sound field between the two speakers simply by controlling the amount of the signal that is fed to each speaker. Of course, this distribution of sound sources is purely an illusion, but it is quite effective in simulating realism during playback.

Besides locating the various instruments at different points

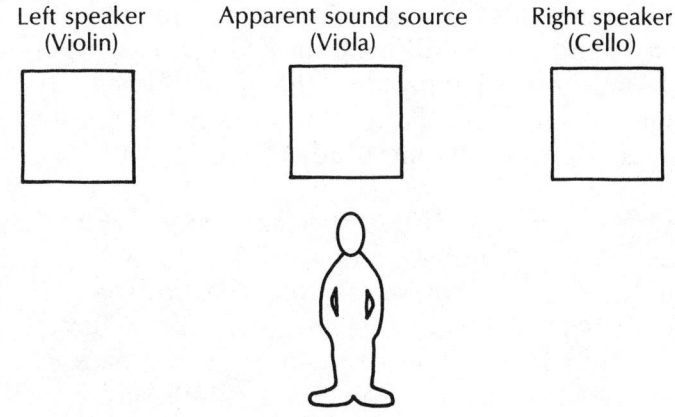

Figure 1-8 *This is the acoustic image of the stereo example described in the text.*

in space, a stereo system heightens realism by placing echoes and reverberations from the original recording environment at different locations.

The effect of stereophonic sound reproduction was such an improvement over straight monophonic sound, that stereo rapidly became the norm. Virtually all modern high fidelity (realistic reproduction) audio systems are stereophonic.

When more than one speaker system is employed, a new factor must be considered. This is the relative phase of the speakers. Suppose we have two speakers wired out of phase with one another. We will assume the same signal is being applied to both speakers. As one speaker pushes outward, compressing the air in front of it, the other speaker is simultaneously pulling inward, creating a slight vacuum in front of it. The opposite actions of the two speakers effectively cancel each other out. The resulting sound waves will be significantly weakened. (Theoretically, they would be canceled out to total silence, but the problem is rarely this extreme in actual practice.)

Of course, in a stereo sound system, the exact same signal is not fed to both speakers, but much of the signal is common to both channels. Returning to our earlier example of the string trio will illustrate the problem for stereo recordings. The three instruments are recorded with the following acoustic locations: violin, far right; viola, middle; cello, far left.

If the stereo speakers are wired out of phase with one another, the sound of the violin and the cello will not be noticeably affected. But the music played on the viola is essentially a monaural signal, so the cancellation from the out of phase speakers will reduce the apparent amplitude of this instrument considerably.

On the other hand, if two speakers are properly connected in phase with one another, and are fed a common signal, the reproduced sound will be reinforced. The two speakers will be working together instead of against each other. Both speakers will be pushing forward, or both will be pulling back.

The two terminals on a speaker's inputs and on an audio amplifier's outputs are usually marked plus (+) and minus (−) to aid in setting up the proper phase relationship between the two speakers in a stereo system. All pluses should be matched up, and all minuses should be similarly connected.

It is important to remember that we are dealing with relative phase here, not absolute phase, as with a battery. The plus and minus designations used here have nothing much to do with electrical positive and negative. No lasting harm will be done if you wire a speaker into a sound system backwards. It is perfectly all right to connect a + output to a − input, as long as you are consistent and do the same thing with every speaker in the system. Obviously, it is more convenient and less likely to result in confusion if you connect + outputs to + inputs and − outputs to − inputs, but this is a matter of convention, not a strict electrical requirement. (Some audiophiles claim that absolute phase is an important factor in accurate sound reproduction, but there appears to be little scientific evidence to back up their claims.)

No other development in the history of audio recording has equalled the impact of the advent of stereo, although digital recording may very well come close. Digital recording is certainly changing the face of the entire field of audio recording. But we still have a few more analog recording media to consider before we get to the digital revolution.

REEL-TO-REEL RECORDERS

Now, let's consider the three major magnetic tape recording formats. The first tape recorders were all of the reel-to-reel or open-reel type. Other than an occasional short-lived experiment, this situation continued into the 1970s.

In the reel-to-reel format, the tape is wound on an open-ended plastic or metal reel. This reel (called the feed reel) is placed on one of two spindles on the tape recorder. A similar, empty reel is placed on the second spindle. This empty reel is called the take-up reel.

The end of the tape from the full reel is manually fed through a series of tape guides, over the tape heads, and finally to a slot in the take-up reel. In playback operation, the take-up spindle is rotated, winding tape onto the take-up reel, as a capstan spins to pull the tape off of the feed reel. Of course, the tape is pulled past the machine's heads in the process. (In the rewind mode, the motor direction is reversed. The tape is wound back onto the feed reel, and pulled off of the take-up reel.)

Obviously, correct tension is absolutely essential to maintain the proper consistent speed of the tape over the heads, and to avoid stretching (or even breaking) the tape. Even a small fluctuation in the tape speed will result in a very noticeable and objectionable degradation of the reproduced sound. Low-frequency speed fluctuations (below about 10 to 15 Hz) are called *wow*, and high-frequency speed fluctuations are known as *flutter*. Both of these names do a good job of describing their audible effects.

A cheap tape recorder with a low quality motor will generally have relatively high wow and flutter levels. High quality tape recorders are designed to reduce such effects as much as possible.

The earliest tape recorders were of the full-track type. The full width of the tape was used to record the signal, as shown in Fig. 1-9. Once the tape was wound entirely onto the take-up reel, it had to be rewound back onto the original feed reel in order to be played again. Of course, this takes time and is a little inconvenient. Besides, the full width of the tape usually isn't needed to record the signal, so tape is wasted.

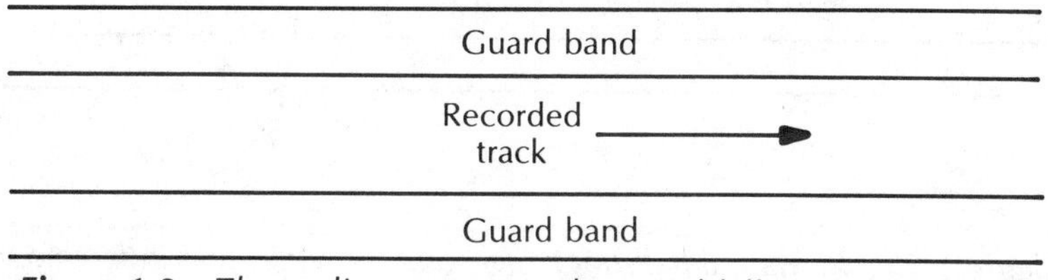

Figure 1-9 *The earliest tape recorders used full-track recording.*

32 The History of Sound Recording and Reproduction

The wider the recorded track, the more space there is available for detail in the recorded signal, and thus, greater fidelity. At reasonably high tape speeds, such as 7½ IPS, the difference between full-track recording and half-track recording is almost negligible, especially in consumer-grade equipment. The increase in convenience and economy more than outweighs the marginal decrease in signal fidelity due to the decreased track width.

On a half-track tape recorder, the signal is placed on just the upper half of the tape's width. When the tape has played through and is wound entirely onto the take-up reel, the positions of the two reels are reversed. The old take-up reel (which currently holds the tape) is the new feed reel and is placed on the first spindle, while the old feed reel (now empty) is placed on the second spindle to serve as the new take-up reel. The tape is pulled past the recorder's heads again, essentially in the opposite direction, to record or play a second program, as illustrated in Fig. 1-10. Notice that by reversing the tape, the old lower track is now the upper track, and vice versa.

Some tape recorders have a special feature known as "auto-reverse." When one side of the tape has finished playing, an electronic sensor causes the motors to reverse direction. The tape starts winding past the heads backwards, eliminating the need to manually change the positions of the two reels. This is usually accomplished by a sensor and a small strip of aluminum tape placed at the end of the tape. Other machines use an optical sensor that detects light through a transparent leader at the end of the tape.

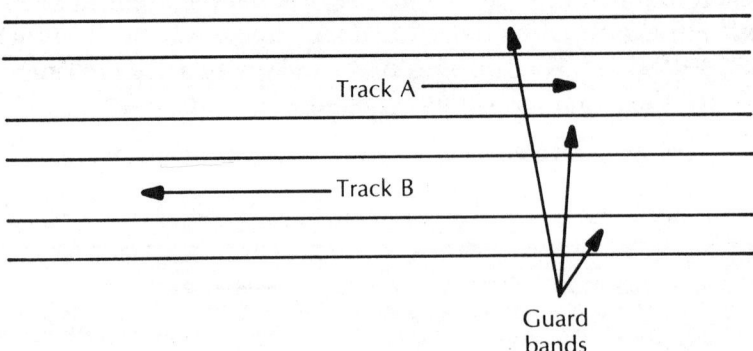

Figure 1-10 *Monaural half-track recording provides two independent programs running in opposite directions.*

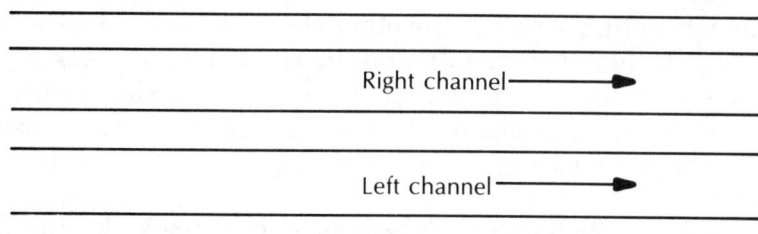

Figure 1-11 *Stereo half-track recording provides two channels of a single program. There is no Side 2.*

When stereo came along, half-track recording became the minimum possible configuration. Full-track recording is inherently limited to monophonic recordings. There is no way to separate the signal information for the right and left channels.

On a stereo half-track tape recorder, one track contains the right channel signal, while the second track contains the left channel. Like the full-track monaural recorders discussed earlier in this section, a half-track stereo tape has no second side. When the tape has been fully wound onto the take-up reel, it must be rewound before it can be replayed. A half-track stereo recording is illustrated in Fig. 1-11.

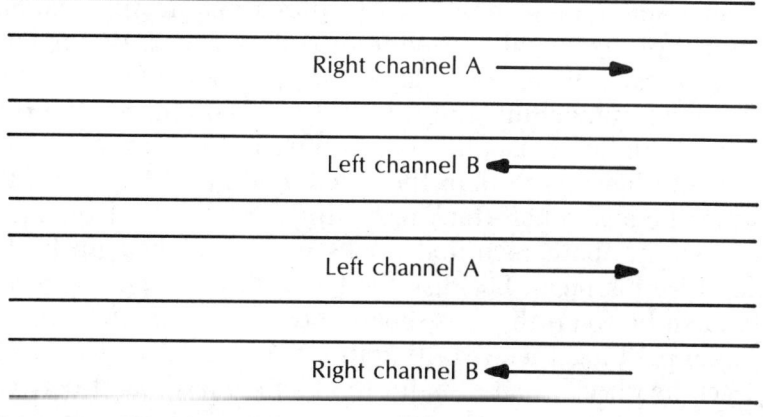

Figure 1-12 *Bidirectional stereo recording can be achieved by splitting the tape into four tracks.*

Bidirectional (two sides or programs) stereo recording can be achieved by dividing the tape into four tracks, as shown in Fig. 1-12. Notice that the tracks of the two programs are interwoven. This helps reduce the problems of crosstalk (part of one channel bleeding over into the other channel, blurring the stereo imaging.) If there is crosstalk and the right channel track is immediately adjacent to the left channel track, the stereo imaging will be blurred. The apparent acoustic location of each instrument will no longer be clear, or perhaps even distinguishable at all.

On the other hand, by interweaving the tracks for the two sides of the tape, as in Fig. 1-12, crosstalk causes some of the opposite program to be played backwards. This might seem to be even more objectionable than a blurred stereo image. Actually, the forward music (which will be much stronger, unless the mistracking and crosstalk are particularly severe) tends to mask the extraneous signal from the unrelated adjacent track. Also, the ear is inclined to ignore the inappropriate crosstalk signal.

Actual listening tests have indicated that crosstalk problems are less audible and objectionable if the tracks are interwoven in this fashion. Therefore, this is the system normally used in most (but not all) four-track stereo reel-to-reel tape recorders. Some machines however, use the upper two tracks for the right and left channels of the first (side 1) program and the lower two tracks for the right and left channels of the second (side 2) program, without any interweaving, as shown in Fig. 1-13.

The track arrangement of the playback machine should match that of the machine used to record the signal on the tape. In some cases, you can use a tape from one type of machine on another type of machine, but this is the exception, not the rule, and sound quality is usually compromised. The playback signal will usually have more than the usual amount of tape hiss. For example, if a stereo half-track recording is played back on a full-track machine, both recorded tracks will be picked up by the single full-track head. Because the head does not (and can not) distinguish between the two separate track signals, the two signals are mixed together, resulting in monophonic playback. The full-track head will also pick up the blank guard band between the two tracks on the half-track recording. The magnetic particles on this portion of the tape are randomly arranged. In other words, this looks like a narrow track of recorded noise. This noise is reproduced right along with the desired recorded signal.

Of course, a bidirectional monophonic half-track recording cannot be played back on a full-track recorder. Both sides will be

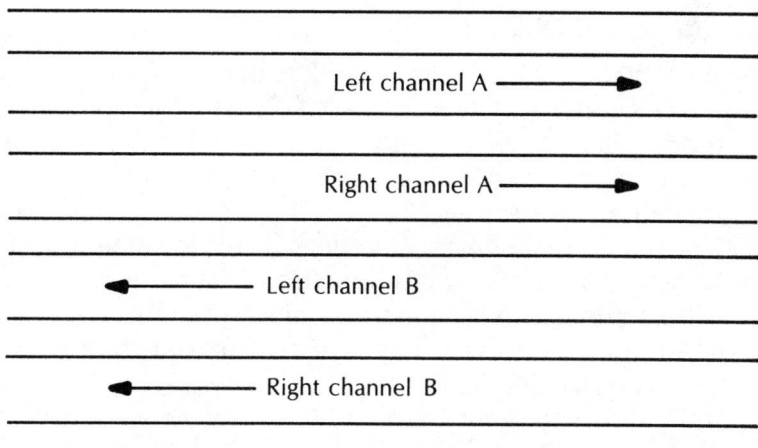

Figure 1-13 *This is an alternate track arrangement sometimes used for four-track stereo recordings.*

heard simultaneously—one of them backwards. Unlike the crosstalk situation discussed earlier, both the forward and backward signals would be at comparable amplitude levels in this case, so the backwards signal will be very audible.

In standard consumer recording equipment, the open reel tape is always one-quarter-inch wide. In professional (studio) tape recorders, one-half-inch and 1-inch wide tapes are often used. The wider tapes are often divided into 8 or 16 tracks for advanced mixing techniques in the recording studio. The principle behind the track divisions are the same as in the consumer machines described here.

The open reel format is mechanically simple and very easy to edit. However, some people find it difficult or, at least, inconvenient to manually thread the tape through the recorder's guides. It is always possible to misthread the tape, which could conceivably result in damage to the tape, or possibly even to the tape recorder itself. Moreover, the open reel leaves the tape's surface open to dust, fingerprints, and other contaminants.

Because of these problems, an enclosed tape format was considered desirable, although it was not practical until the late 1960s. There were a number of different designs attempted over the years, but only two achieved any significant degree of popularity. One of them, however, has already died into obsolescence.

THE 8-TRACK CARTRIDGE

In the late 1960s and the 1970s 8-track cartridge tapes were quite popular, especially in car stereo systems. An 8-track cartridge has a rectangular plastic housing. A simplified diagram of an 8-track cartridge is shown in Fig. 1-14. Notice that the tape is entirely contained within the housing. An opening in one end of the cartridge permits the tape recorder's heads to come in contact with the tape. The tape is wound into a special endless loop, and the ends are spliced together with a piece of aluminum. When this aluminum band passes a special sensor near the recorder's head, an electromechanical switch is activated. The head is automatically moved to play the next set of tracks on the tape.

Since the format is called an 8-track cartridge, it obviously follows that the tape width is divided into eight recording areas, or tracks. Since this is a stereophonic recording medium, two tracks are required for each program. Therefore, an 8-track cartridge contains four individual stereo programs, which are played sequentially.

The arrangement of tracks on an 8-track tape is illustrated in Fig. 1-15. Notice that all eight tracks run in the same direction. The motion of the tape in an 8-track cartridge can never reverse direction.

At the end of the fourth program, the head is returned to the first program position and starts over. The tape plays continuously. This continues as long as the cartridge is plugged into the machine. Another sensor shuts the recorder off when the tape is

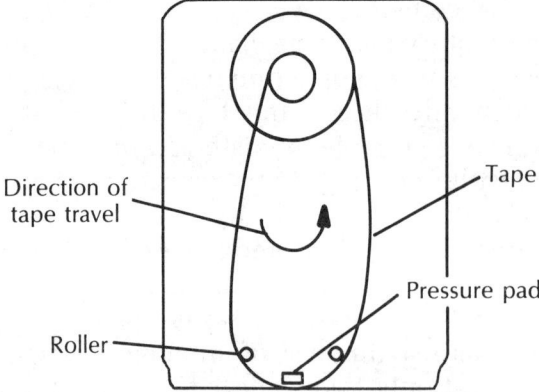

Figure 1-14 *The 8-track cartridge tape was popular in the late 1960s and the 1970s.*

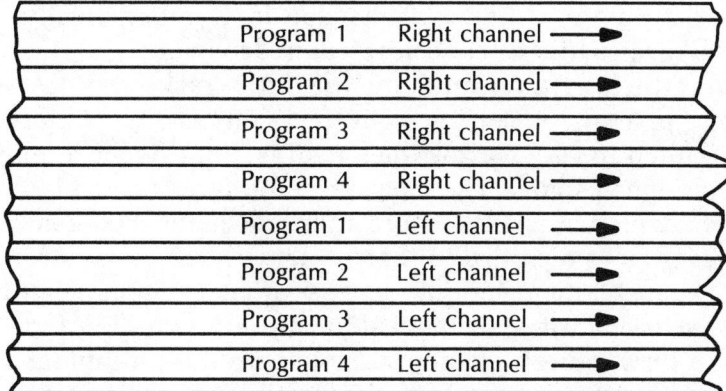

Figure 1-15 *The tracks on an 8-track tape are arranged like this.*

removed. Some deluxe 8-track tape players featured automatic shut off at the end of the current program, or after all four programs had been played.

It is absolutely impossible to rewind the tape in an 8-track cartridge due to the endless loop design. Tape can be wound on the endless loop in only a single direction. If you tried to rewind an 8-track tape, it would very quickly become hopelessly jammed.

On most 8-track tape players a manual switch of some kind is provided to manually select one of the four adjacent programs on the tape, and sometimes a fast forward control is included; although fast forward in an 8-track cartridge system isn't really very fast. Typically, the fast forward speed is only about twice the ordinary playback speed. Once again, attempts to fastwind the tape on the endless loop within the cartridge would inevitably result in extreme jamming problems.

Most consumer 8-track cartridge machines were playback units only, although a few home 8-track cartridge recorders were successfully marketed. However, this tape format never became particularly popular as a home recording medium. The main reason for this is probably that it was so difficult to cue up a tape to a specific desired point without rewind or true fast forward capabilities.

Eight-track cartridge players were popular for background music, and especially in car stereo systems, because they can run continuously while unattended. In a car stereo system, a driver could just plug in a tape cartridge and let it automatically play over and over while he concentrated on driving.

Nevertheless, the limitations of the 8-track cartridge soon became apparent. Squeezing eight tracks onto a one-quarter inch tape resulted in very narrow track widths, limiting the fidelity of the reproduced signal. The tape speed for 8-track cartridges was 3¾ IPS, which was little more than adequate. Frequency response for tapes recorded in this format was fair at best.

The most critical and inescapable problems with the 8-track cartridge format were mechanical in nature. Rewinding and even true fast forwarding were not possible with the continuous loop arrangement of the tape. The endless loop cartridge was relatively delicate, and had to be perfectly assembled or it wouldn't operate properly. Jamming was an all too frequent problem with 8-track cartridges. Finally, because of the complicated continuous loop mechanism, editing the tape in an 8-track cartridge was virtually impossible.

Eventually, almost everyone switched over to audio cassettes (discussed in the next section of this chapter), and the 8-track cartridge quietly died off. Some 8-track players and recorders are still in use, but I don't think there have been any commercially recorded 8-track cartridges released in years.

THE AUDIO CASSETTE

One of the most popular media for recording music was never really intended for that purpose when it was designed. The Phillips Company originally created the audio cassette for use in dictation systems. However, the inherent convenience of this format was so appealing that considerable work was done until the fidelity obtained with a cassette tape recorder rivals that of a semiprofessional reel-to-reel recorder. At the time of this writing, audio cassettes are the number one recording medium on the market today, although CDs appear to be catching up.

The basic design of the audio cassette is illustrated in Fig. 1-16. As you can see, the cassette is basically a miniature reel-to-reel system, with the tape prethreaded, and both reels enclosed in a single plastic housing. To play (or record) the tape, the user only has to insert the cassette into an access well and push the appropriate buttons on his machine. The user never actually touches the tape. In fact, like any magnetic tape, it is a

The Audio Cassette 39

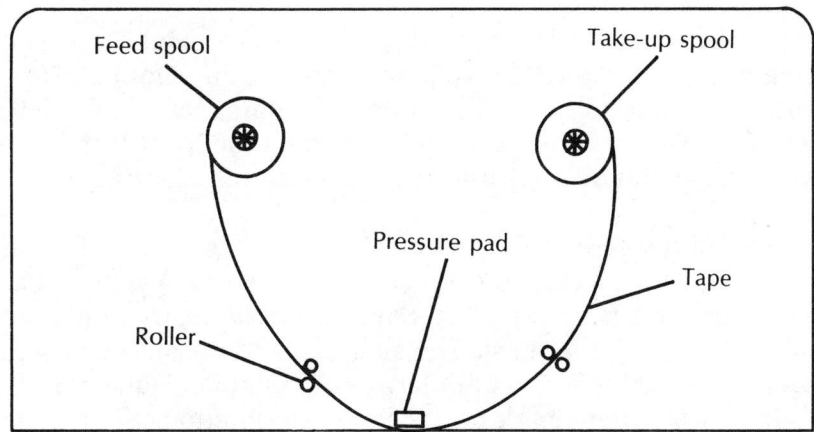

Figure 1-16 *The audio cassette resembles a miniature reel-to-reel system, enclosed in a protective plastic housing.*

good idea never to touch the tape contained in an audio cassette. Dirt and natural oils from your fingertips can build up as contamination on the tape surface. This contamination will interfere with the tape recorder's heads ability to accurately pick up the recorded signal on the tape.

Because of the enclosed nature of the cassette housing, editing this type of tape is difficult, though it is possible with patience and a steady hand. The problem here, of course, is the narrow width of the tape. For editing, several feet of the tape must be pulled out of the protective cassette housing. It can all too easily get twisted, stretched, broken, or dirty. Then, after the editing process, the tape must be carefully wound back onto the hubs within the cassette housing. If the tape is improperly loaded into the cassette, jamming will result.

Even more so than with open reel tapes, care must be taken to minimize touching (and contaminating) the oxide side (recording surface) of the tape during editing (or any other time, of course). Some experts recommend wearing special lint-free gloves while editing, especially when working with cassette tapes. Dedicated editing and splicing blocks for one-eighth-inch cassette tapes are available from a number of sources, including Radio Shack.

The tape path in an audio cassette is far simpler mechanically than in an 8-track cartridge. The tape path is simple and straightforward, basically resembling the system used with reel-to-reel tape recorders. There is no complex and critical endless loop to worry about in an audio cassette. As a result, tape jamming is generally much less of a problem with this format. However, audio cassettes are still prone to jamming, especially with a cheap, poorly assembled housing. Dirt on the recorder's heads and capstan can also significantly increase the chances of tape jamming.

One of the audio cassette's most attractive features is its conveniently compact size. A cassette tape or two can easily be carried in an average shirt pocket. Several will fit neatly into a typical coat pocket or a purse. The small size of the audio cassette was a major factor in the popularity of "Walkman" tape players. (Walkman is a trademark of the Sony Corporation. Similar devices from other manufacturers are usually called personal stereos or personal tape players.) In the car, several audio cassettes can conveniently be stashed in the glove compartment.

Unfortunately, the small size is probably also the chief drawback of the format. The tape is only one-eighth-inch wide, and creeps past the recorder's heads at only 1⅞ IPS. This puts some limits on the inherent fidelity possible with this recording system. Advanced technology has come a long way in overcoming the inherent limitations of the audio cassette.

Originally, the audio cassette was designed for voice rather than music applications, so the recordings were made in mono. The tape was divided into two tracks, one for each side (or program), as illustrated in Fig. 1-17.

Of course, once audio cassettes were used for music recording, stereo was considered essential. The tape width was then further divided into four tracks, as shown in Fig. 1-18. Notice that the two channels of each program are placed side by side, rather than interwoven, as in the four-track reel-to-reel tape recorders discussed earlier in this chapter. This was done to preserve compatibility with pre-existing monaural cassettes. A cassette recorded on a monaural machine can be played back on

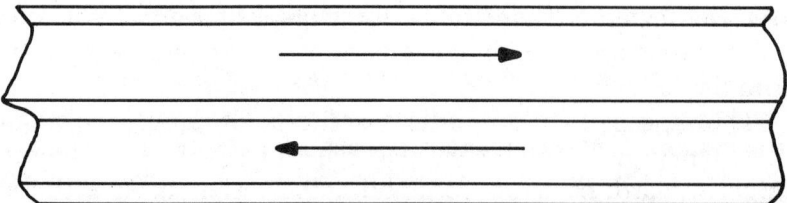

Figure 1-17 *The first audio cassettes were designed to hold two monophonic programs.*

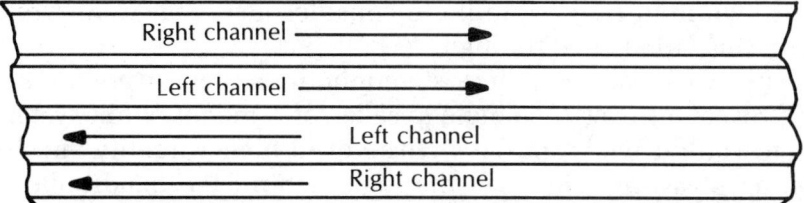

Figure 1-18 *To record stereo programs, the cassette tape is divided into four tracks.*

stereophonic equipment, and a cassette tape recorded on a stereo tape recorder can be played back (in mono, of course) on a monaural machine. This compatibility was a requirement of Phillips's licensing agreements with other manufacturers for the audio cassette. This enforced compatibility was probably a major factor in establishing the popularity of this tape format. Consumers did not have to choose between competing and incompatible systems.

There is one particularly nice feature available on audio cassettes. There is a small tab on the back of the cassette housing. As long as this tab is in place, the tape may be recorded on. But if this tab is broken off, a sensing mechanism in all cassette recorders prevents the record button from working. The tape can be played back, but not recorded. This prevents accidental erasure. If the user later changes his mind and wants to erase a tape with the tab broken out, a piece of ordinary cellophane tape over the hole will fool the recorder's sensing mechanism, allowing the record function to work just as if the tab had never been broken out.

Some more advanced cassette recorders sense the presence or absence of a second tab. If this second tab is broken out, the recorder's circuitry adjusts itself to accommodate a different tape formulation, such as chromium dioxide (CrO_2). If the missing tab hole is not sensed (either it's not broken off or the cassette housing has no removable tab at this point), the machine assumes the tape used is of the standard (ferric oxide (Fe_2O_3)) formulation. Other cassette decks have a manual switch, permitting the user to adjust the circuitry for the proper tape formulation.

THE QUADROPHONIC FIASCO

In the 1970s there was an attempt at a major revolution in the audio field which flopped, largely due to clumsy marketing. It is worthwhile to take a minute to consider just what went wrong with this failed breakthrough.

If two speakers, as in a stereophonic system, were a major improvement over the single speaker of a monophonic system, then wouldn't four speakers create even higher fidelity and listening pleasure? This was the logic behind the quadrophonic sound system.

In a quadrophonic system, there are two front speakers and two rear speakers. The listener is positioned in the middle, as shown in Fig. 1-19. This arrangement surrounded the listener with sound, placing him in the middle of a three-dimensional sound field.

The four speakers in a quadrophonic system are labeled as follows: right front, left front, right rear, and left rear.

The right front and left front speakers correspond more or less with the right and left speakers of an ordinary stereophonic system. When ideally used, the listener was placed in an illusionary concert hall. The instruments appeared to be located in front of him, between the right front and left front speakers. The right rear and left rear speakers produced mainly ambient sound, echoes, and reverberations from the original recording environment, which overpowered the more acoustically restrictive char-

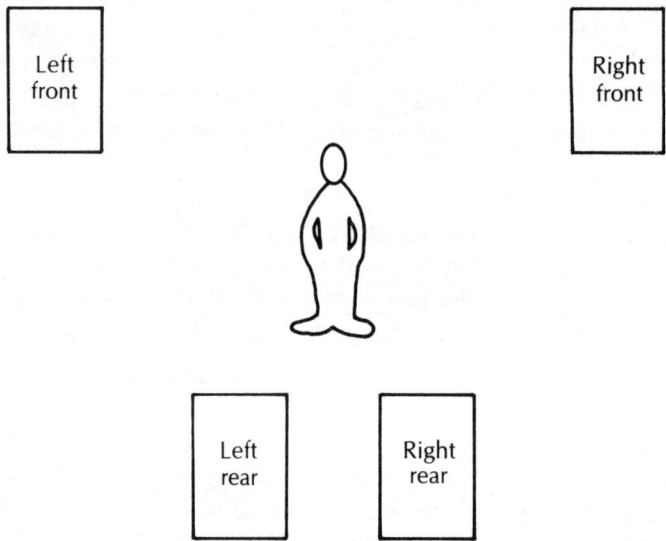

Figure 1-19 In a quadrophonic sound system, the listener is placed in the center of four speakers.

acteristics of most listening rooms. When well done, the effect could be startingly realistic. Occasionally, for special effects, a sound might be located behind the listener; for example, an off-stage chorus in an opera might be placed between the rear speakers.

I remember a very effective quadrophonic recording of Tchaikovsky's *1812 Overture*. For the first part of the recording, all the instruments were located in front of the listener, as in a standard orchestral recording. But when it came time for the climactic cannons, they were fired behind the listener. I don't remember anyone who didn't jump the first time they heard this. It was very effective emotionally, and suited the piece well.

Too often, however, producers of quadrophonic recordings didn't trust listeners to appreciate the subtle, but definite increase in sonic realism provided by a quadrophonic system. Instead, these producers assaulted the listener with extravagant (and unrealistic) "ping-pong" effects, moving the apparent location of instruments around the listener for no good (musically sound) reason. This kind of thing can sound "nifty" the first time you hear it, but the novelty quickly wears off, and the listener realizes its pure gimmicky hokum and scarcely worth the extra cost and space. Since a quadrophonic system, by definition, requires four speakers, it would obviously be terribly cramped in a small room or a studio apartment.

Other producers placed some of the musical instruments behind the listener via the rear speakers. The effect, while strikingly realistic, was rather disconcerting to most listeners. Instead of sitting back in a choice seat in a concert hall or auditorium, they were thrust to the middle of the stage with the orchestra or band. This certainly isn't the way most of us normally listen to music.

The bad production techniques used on so many quadrophonic recordings rapidly earned the system the undeserved reputation of being nothing more than a gimmick. Many consumers rapidly lost interest in quad after hearing a few poor demos.

What really killed quad in the marketplace was the total lack of standardization. There were several competing and noncompatible quadrophonic systems vying for consumers. It was obvious that at least some of these systems would soon become obsolete. If not enough equipment that could play quadrophonic system A was sold, it would not be profitable for recording companies to release records in that format, and everyone who bought system A would have wasted their money.

So the public as a whole sat back and waited to see which system would end up dominating the quadrophonic arena. Sales for all of the competing systems were too low to justify many releases from the record companies, so the public bought even fewer quadrophonic systems because of the shortage of recordings. This vicious cycle rapidly killed off all of the various competing quadrophonic systems, and no one won. Today, quad is every bit as dead as the 78-rpm disc.

It's really a shame that quad never really had a chance to take off and find its niche in the audio marketplace. It was a good idea, and when well done, a good quadrophonic recording could really heighten the realism and listening pleasure.

One good thing came of the whole quadrophonic fiasco. Manufacturers of audio equipment learned that in a format war, everybody can lose. With later important developments, manufacturers tended to be somewhat more inclined to get together and agree on format standards before turning out something totally new. Both the CD and DAT grew out of extensive multimanufacturer conferences to hammer out an industry standard.

A VIDEO DIVERSION

In this book, we are concerned with audio recording, but it is worthwhile to take a moment to consider video recording.

One of the hottest products of the late 1970s and the 1980s was the VCR (video cassette recorder). A VCR allows the user to record television programs—both the video and the audio signals are recorded.

It is much more difficult to record video signals than audio signals. Video signals are inherently much more complex, and much more information must be recorded. To fit enough information onto the tape, it must move past the recording and playback heads at very high speeds. Unfortunately, this isn't very economical. It would take a very long tape to record even a short program. No one would buy a VCR with a maximum recording time of 5 or 10 minutes.

The manufacturers of VCRs came up with a very clever solution to this dilemma. First, a one-half-inch wide tape was enclosed in a cassette housing. When inserted into the VCR machine, some of the tape was pulled out of the cassette and wrapped partway around the recorder's head drum, as shown in Fig. 1-20. Two or four separate heads are mounted on the drum. As the tape is slowly pulled past the drum, it is revolving at a very high speed. Because the recorder's heads are moving with respect to the tape, the effective tape-to-head speed is very high. The heads are angled to record narrow diagonal tracks along the width of the tape, as illustrated in Fig. 1-21. In this way, large

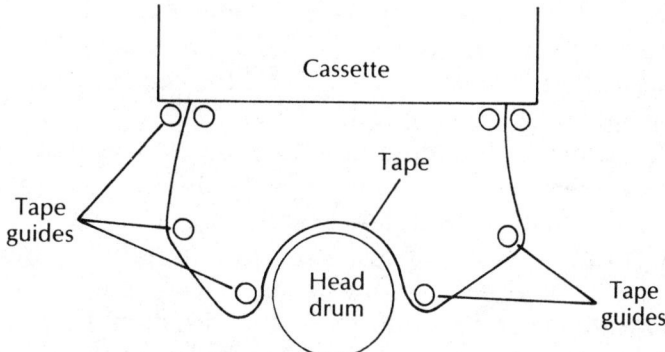

Figure 1-20 A VCR uses a revolving head drum with the tape wound partially around the drum.

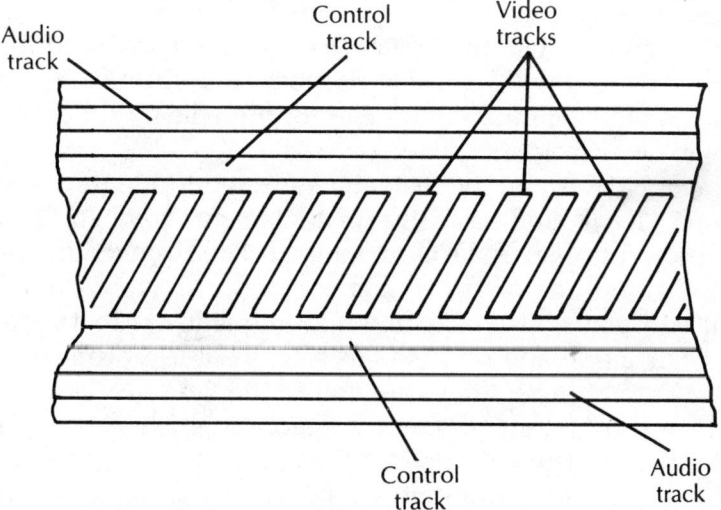

Figure 1-21 A VCR records the signal in diagonal tracks across the width of the tape.

quantities of recorded information are squeezed onto a relatively small amount of tape.

When we get to digital audio recording, we will see that digital data has similar high information content requirements. DAT recorders also use rotating drum heads, like VCRs. If it wasn't for the VCR, we probably wouldn't have DAT today.

PULSE CODE MODULATION (PCM)

Often in communications (recording and radio transmissions), some form of modulation is used. In any modulation system, there are two important types of signals. The *carrier* is a continuous reference signal of some sort, and the *program* is the varying signal to be recorded or transmitted.

In the process of modulation, the program signal is imposed onto the carrier signal in some way. In the receiver (playback unit), the original, unmodulated carrier signal is recreated, and the deviations from its normal, constant condition are used to reproduce the original program signal.

There are many different types of modulation. In analog systems, the two most common types of modulation are AM (amplitude modulation) and FM (frequency modulation).

In an AM system, the instantaneous amplitude (level) of the program signal modulates the instantaneous amplitude (level) of the carrier signal. The stronger the instantaneous amplitude of the program signal, the higher the instantaneous amplitude of the carrier signal will be at that moment. Amplitude modulation is the simplest modulation system to implement, but it is quite prone to noise from interference.

In an FM system, the instantaneous amplitude (level) of the program signal modulates the instantaneous frequency, rather than the amplitude, of the carrier signal. The stronger the instantaneous amplitude of the program signal, the more the carrier signal will deviate from its nominal frequency at that moment. Frequency modulation tends to be somewhat less prone to interference than amplitude modulation (although noise can still be a problem), but it generally requires more sophisticated circuitry.

In digital systems, some form of PCM (pulse code modulation) is used. Most forms of PCM can also be used in strictly analog applications.

In any PCM system, the carrier signal is always in the form of a rectangular or square wave. In other words, the carrier is a stream of pulses. The carrier pulses can be modulated by the program signal in a variety of ways. For example, the number or rate of pulses may be varied. In a PAM (pulse amplitude modulation), the height (amplitude) of the pulses follows the changes in the amplitude of the program signal. Another popular form of PCM is PWM (pulse width modulation). Here the width, or length, of each pulse is dependent on the instantaneous amplitude of the program signal. Some form of PCM is used in any type of digital recording.

THE COMPACT DISC

The compact disc (CD) was introduced in the 1980s. It was the first practical consumer-level digital recording medium, although on the consumer level, it is a playback only medium.

A CD is a metallic disc about 5¼ inches in diameter. It has a moderately large center hole. When held under a strong light source, the CD's surface has a kind of "rainbow" reflective quality. Etched into its surface (under a protective plastic coating) are millions of microscopic pits and islands representing digital bits. The digital bits thus stored are a modulated representation of the original (analog) audio program.

Some CDs contain programs that were originally recorded on analog tape recorders, and later transferred over to the digital realm. Others start out with a digital tape recorder and remain entirely in the digital realm until the consumer plays the disc. Of course, to hear the music, it must be in analog form.

A unique feature of the CD is that it is played without anything in the playback mechanism actually touching the recorded surface. There is no wear and tear involved in playing a CD. A low-power laser (which we can consider as just a highly focused light beam) is aimed at the disc's surface, and a set of sensors measure the light beams reflected back from the surface. The pits and islands return different reflection patterns. The playback circuitry reads the differences between them and decodes the modulated digital pulses.

Another unusual feature of the CD is that it is played from the inside out. The recorded data starts near the center of the disc and moves outward towards the rim. The disc spins at a very high rate, which varies depending on whether the laser beam is reading a track near the center or near the outer edge of the CD.

The CD certainly earns the compact part of its name. While

a CD is only 5¼ inches in diameter, it can easily hold more music than a 12-inch LP. A CD can hold up to 72 minutes of music. (Some discs have even longer playing times, thanks to improved recording techniques.)

Compact discs feature very clear, noise-free sound, long playing times, and extremely good durability. They aren't quite indestructable (despite some early claims), but they're reasonably close to it. To damage a CD you pretty much have to work at it. It can be done, but not very easily.

After a fairly slow start, CDs caught on with the general public, then rapidly took off in the marketplace. Virtually all new recordings are now being offered on CD, and most major record labels are rereleasing their back catalogs in this exciting new format.

For all intents and purposes, the CD is rapidly taking over the place previously held by the vinyl record album. Go into any record store today and you'll find less shelf space devoted to LPs, and much, much more devoted to CDs.

Currently, a considerable amount of research is going into the development of an erasable, user-recordable CD. Compact discs are sometimes referred to as "optical discs" because they are read by light, rather than any mechanical contact. True digital data can also be stored on such a disc and used with a computer system. The technical details of the CD format will be discussed in more detail in Chapter 4.

HERE COMES DAT

Now, as we are entering the 1990s, DAT (digital audio tape) is slowly showing up on dealers' shelves, after several years of promises. There have been a number of legal problems delaying the U.S. marketing of these exciting new machines, but these finally seem to be resolved (or close to it). The legal battles of DAT will be discussed in Chapter 6.

While the CD is the digital equivalent of the vinyl LP, DAT could be considered the digital equivalent of the audio cassette. In fact, DATs are housed in cassettes that are similar to (but unmistakably incompatible with) their analog cousins.

The next chapter will go into a bit more detail about the technical aspects of analog sound recording. We will then delve into the fascinating and powerful world of digital recording in Chapter 3.

❖ 2
Analog Sound Recording

DIGITAL SOUND RECORDING IS FAIRLY COMPLEX. TO UNDERSTAND how it works, it is very helpful to first have a basic understanding of how sound is recorded in the analog realm. That is the purpose of this chapter. In the following pages we will take a close look at the two most widely used types of analog recording media—the vinyl disc and magnetic tape. We will not bother with obsolete techniques such as the Edison cylinder, with its vertically cut grooves, or the wire recorder.

For magnetic tape recording, one general explanation covers all the formats. The same principles are used in reel-to-reel, cassette, and 8-track cartridge recorders. The only significant difference between these three competing tape formats is in the mechanical threading and storage of the tape. The same basic types of heads and electronic circuitry are used in all three of the major magnetic recording formats. The signal is recorded onto the tape in exactly the same way in each case.

THE MODERN PHONOGRAPH

A phonograph record, as you recall, is a vinyl disc with tiny lateral grooves cut into its surface along a continuous spiral path, starting from near the outer rim of the disc and working its way inward towards the center label. For playback, this disc is placed on a platter, or turntable, which rotates at a fixed speed. A pivoted tonearm is placed over the disc. Mounted on the underside of this tonearm is a sapphire or diamond needle, or stylus, which rides in the grooves. The laterally cut grooves force the stylus to move back and forth, and these variations from a straight (or

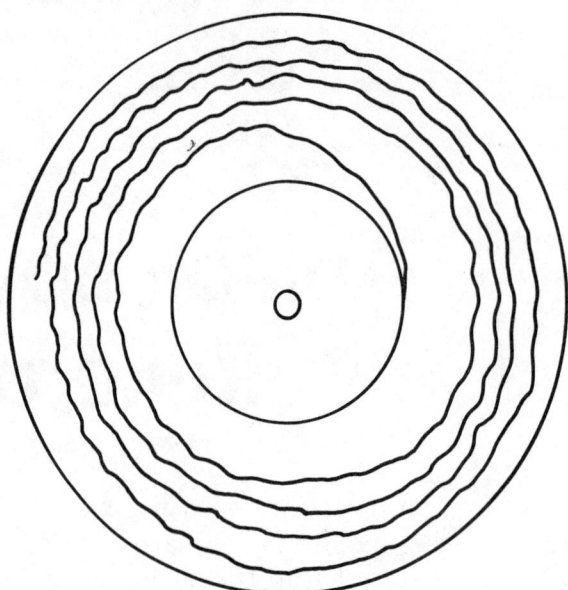

Figure 2-1 The signal on a disc recording is in the form of undulating spiral grooves cut into the surface of the disc.

actually smooth spiral) path are used to generate an electrical signal, which can be amplified and fed to the system's loudspeakers.

Figure 2-1 illustrates the way these undulating grooves are cut into a record disc. The grooves in this drawing are very greatly exaggerated for clarity. On a real record, the grooves are nearly microscopic and very closely placed. The further a groove varies from its nominal, no-signal path, the higher the instantaneous amplitude of the recorded signal.

The device that does the actual work of converting the varying mechanical motion of the stylus into an appropriate electrical signal is called the cartridge. Once the signal is in proper electrical form, it can be amplified and otherwise manipulated by the reproduction equipment in the sound system. For example, the system frequency response can be varied somewhat to account for the physical characteristics of the listening room, and for personal taste. The playback signals remain in purely electrical form until they are converted back into mechanical (sound) energy by the loudspeakers.

While there are a number of variants, there are two basic types of cartridges commonly used in modern record playing systems. These are ceramic cartridges, and magnetic cartridges. We will consider each of these cartridge types separately.

Both of these basic cartridge types are fairly sophisticated when compared to the styluses used in the early Edison and Berliner phonograph machines. For the first couple of decades of the phonograph's existence, such odds and ends as cactus needles and sewing needles were pressed into service as record playing styluses. In modern equipment the stylus itself is usually a small chip of diamond. Some cheaper units are sapphire styluses, but these don't last as long as the diamond type, and are more likely to damage the grooves of the record being played.

Ceramic cartridges

A ceramic cartridge works by a principle known as the piezoelectric effect. The physical motion of the stylus places a mechanical stress along one axis of a piece of ceramic material. This causes a proportionate electrical stress to be generated across the opposite axis of the ceramic crystal in a different plane.

Ceramic cartridges offer a number of desirable advantages. They are generally inexpensive and relatively sturdy, and they produce a fairly strong output signal. Ceramic cartridges are generally used to drive an audio amplifier directly. No special external preamplifier stage is usually needed with a ceramic cartridge.

These significant advantages of the ceramic cartridge, however, are offset by certain important disadvantages. The primary disadvantages of the ceramic cartridge include a relatively poor frequency response and a high tracking force. The tracking force of a cartridge is the amount of pressure (or the effective weight) exerted by the stylus on the record's grooves. Obviously, a high tracking force will tend to wear down the grooves (thus rendering the record useless) faster than a lighter tracking force.

It is important to realize than any record disc deteriorates a little each and every time it is played. This is especially true when a ceramic cartridge is used to play the disc. It should be perfectly obvious when you think about it—the higher the tracking force, the greater the pressure exerted on the relatively thin groove walls. Since there is more pressure, the groove walls are more likely to be chipped or scratched as the record is being played.

The pressure from the stylus tends to flatten out the variations in the groove walls. Theoretically, if a record was played enough times, the groove walls would be completely flattened out. The recorded music would effectively be erased. In practical terms, because of the substantial increase in surface noise, the record would probably be discarded long before this point is reached.

Tracking force is normally measured in grams. Most ceramic cartridges exert a tracking force of about 5 grams. Some better quality ceramic cartridges may feature a tracking force as low as 3 grams. "El cheapo" cartridges can be expected to exert tracking forces considerably greater than the nominal 5-gram average usually stated for ceramic cartridges as a class.

Ceramic cartridges are found in most inexpensive record players, especially the all-in-one types. Better systems generally use separate turntables with magnetic cartridges.

Magnetic cartridges

From the 1960s on, most better quality turntables used magnetic cartridges instead of less expensive ceramic cartridges. In a magnetic cartridge, the stylus is physically attached to a small piece of magnetic material that is moved between two small coils, as shown in Fig. 2-2. The magnet's mechanical motion follows that of the stylus. Of course, a moving magnet implies that the magnetic field surrounding the magnet must also be moving in perfect step with the magnet itself. This moving magnetic field induces an ac voltage into the coils.

What we have here is the principle of induction, one of the primary theories of electronics. If a coil is moved through a suf-

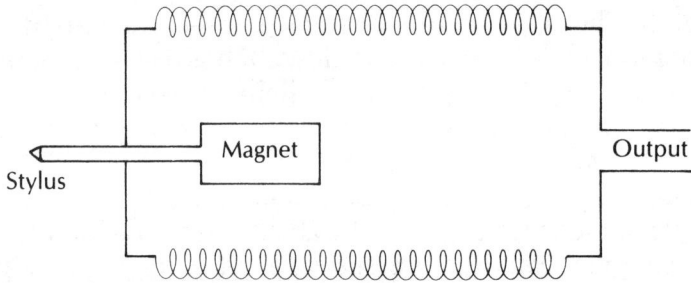

Figure 2-2 *A magnetic cartridge is somewhat more sophisticated than a ceramic cartridge.*

ficiently strong magnetic field (or if a moving magnetic field moves across a stationary coil), a proportional voltage will be induced into the coil. Similarly, if an ac voltage is fed through a coil, the coil will be surrounded by a proportional magnetic field.

So, when a magnetic cartridge is used, the stylus motion is converted into a moving magnetic field that induces a proportional ac voltage into the cartridge's coils. This ac voltage is a reconstruction of the original electrical signal encoded into the grooves of the disc (ignoring the effects of any distortion or noise that might have been added to the signal at any point along the audio recording and reproduction chain).

Magnetic cartridges typically offer a much better frequency response than ceramic cartridges. Magnetic cartridges generally offer the most improvement in the high-frequency response. It makes sense that a lighter stylus can respond faster and more accurately to rapid changes in the groove direction because there is less mass to be moved.

The tracking force of a magnetic cartridge is also far lower than that of a ceramic cartridge. A typical range of tracking forces for magnetic cartridges is about 0.75 grams up to a maximum of approximately 3 grams. Most low- to medium-priced magnetic cartridges probably have a tracking force of about 1 to 1.5 grams or so.

As you can see, the magnetic cartridge is definitely superior in its potential performance to the ceramic cartridge. However (and there always seems to be a however), magnetic cartridges tend to be considerably more expensive than ceramic cartridges. They are also rather delicate and easily damaged if handled carelessly. A magnetic cartridge can be permanently damaged if it is dropped. A ceramic cartridge, while still relatively fragile (because of its tiny component parts) can survive much rougher handling than a magnetic cartridge. For a young child's first record player, or for a portable record player, a magnetic cartridge would probably be a big mistake.

Another significant disadvantage of a magnetic cartridge is that it puts out a relatively weak electrical output signal. Some sort of preamplification stage is almost always required with a magnetic cartridge before the electrical signal is strong enough to successfully drive a regular audio amplifier.

Cutting the grooves

The grooves in a record disc can theoretically be cut at virtually any angle with respect to the surface of the disc. (In practice, there are some reasonable limits, of course.) For example, the stylus could be forced by the groove undulations to move laterally (from side to side) or vertically (up and down, as in the early cylinders). In either case, the output from the cartridge would be exactly the same. In practice, vertical groove records are fairly difficult to reproduce and manufacture in quantity, so lateral motion is the norm, at least for monaural recordings.

Now, suppose we had a groove with both lateral and vertical modulations. Since the two motions are in completely different planes (at 90-degree angles to each other), two separate signals can be detected by a single stylus, even if both motions take place simultaneously. This gives us a full stereophonic (two channel) recording.

Because of the inherent difficulty involved in cutting and copying variable depth grooves, and to ensure compatibility with existing monaural equipment, in practical stereo discs, both sets of groove undulations are rotated with respect to the disc so that they are both at 45-degree angles to the surface of the disc, as illustrated in Fig. 2-3. This is the way stereo signals are recorded onto standard vinyl discs. A specially designed cartridge is required to detect and reproduce the stereo effect. A monaural cartridge would simply combine both signals into a

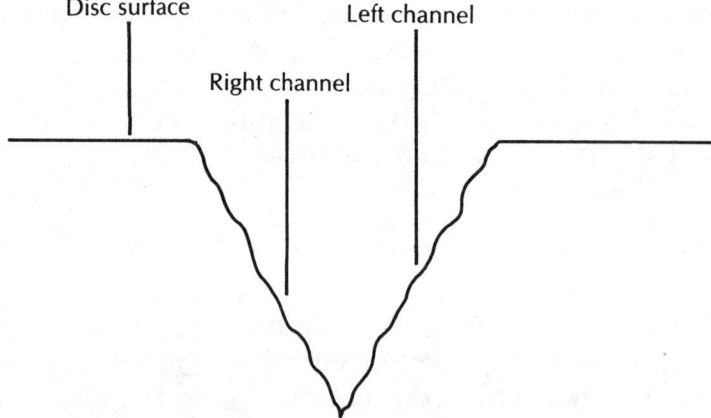

Figure 2-3 *For stereo discs, the grooves for the two channels are placed at opposing 45-degree angles from the surface of the disc.*

single output. All separation between the two stereo channels would be completely lost. If fed through a stereo amplifier, the right and left speakers will both be fed the same exact signal.

A stereo cartridge, on the other hand, is specifically designed to translate one set of stylus motions (in one plane) into information for reproducing the right channel, and the other set of stylus motions (in a second plane) into information for reproducing the left channel.

In any practical audio equipment, there is inevitably some leakage, or crosstalk, between the two channels. That is, if a disc was recorded with only right channel groove modulations, and nothing in the left channel groove, there will still be some (albeit, probably weak) signal showing up in the left channel output too. Crosstalk is largely a function of the cartridge design. Better quality cartridges usually have lower crosstalk ratings.

Bear in mind, however, that crosstalk can show up at almost any point in the audio chain. It can actually be recorded onto the disc, or it can show up in one of the stereo amplification stages.

Equalization

Because of inherent frequency response limitations in recording and playback of discs, certain frequency components in the signal are normally pre-emphasized during recording, and de-emphasizes in playback. This process is known as equalization. Simply put, *equalization* is a deliberate unequal amplification of different frequencies. Low frequencies and high frequencies are given more amplification, or boost, than midrange frequencies. This helps flatten out the record's frequency response.

The standard equalization curve used on virtually all modern record discs is defined by an industry-wide organization known as the RIAA (Recording Industry Association of America). Not surprisingly, the standard equalization curve used for discs is known as RIAA equalization. The concept of frequency equalization will be discussed in a little more detail in the following section on tape recording.

Wow and flutter

The motor driving a turntable should be very accurate and stable. Obviously, it must turn the platter at the correct speed or the music will be too fast and too high pitched or too slow and too low pitched. Except for audiophiles with highly trained ears, if the turntable speed is off by a few percent, it probably won't make any noticeable difference. More important than the precise speed is that the turntable speed be very stable and constant. If the turntable's rotation periodically speeds up or slows down, the effect will be very audible and quite objectionable.

Low-frequency fluctuations in speed are called *wow*. This is because, if the effect is exaggerated, the reproduced sound has a "wow-wow-wow" quality to it. With wow, the speed fluctuations are slow enough that the ear can catch each slowing down and speeding up.

If the speed fluctuates at a higher rate, the ear won't be able to directly recognize the up and down effect, but the reproduced sound will be modulated by the speed fluctuations, producing a bubbly, fluttering quality known as *flutter*.

Specifications for turntables always include wow and flutter ratings. These are probably the most significant specifications for the turntable itself.

MAGNETIC TAPE RECORDING

All three of the major tape recording formats (reel-to-reel, cassette, and 8-track cartridge) operate according to the same basic principles. The differences between the three tape formats are mechanical, not operational. For this reason, in the following discussion, we will assume that we are dealing with a reel-to-reel tape recorder like the one shown in Fig. 2-4.

Modern recording tape is simply a long strip of plastic that has been coated with tiny magnetically charged particles, suspended in a glue-like substance known as the binder. The plastic strip is called the base, or backing.

Usually, the magnetic particles used in recording tape are some form of ferric oxide (Fe_2O_3), but newer technology often uses other substances, such as chromium dioxide (CrO_2), to make improved tapes. Different formulations are especially popular in the audio cassette format. Regardless of the specific formulation employed, the basic operating principles remain the same.

The record head

On a blank (unrecorded) tape, the magnetic particles are arranged haphazardly, as illustrated in Fig. 2-5. If a strong mag-

Magnetic Tape Recording 57

Figure 2-4 This is a typical reel-to-reel tape recorder.

Figure 2-5 On a blank tape, the magnetic particles are randomly arranged.

netic field is placed across the tape, the magnetic particles will line up in that magnetic field, as shown in Fig. 2-6. However, lining up all of the particles along a length of tape in one constant direction doesn't really accomplish much of anything useful.

In a practical tape recorder system, the record head generates a magnetic field that is placed across the tape. The record

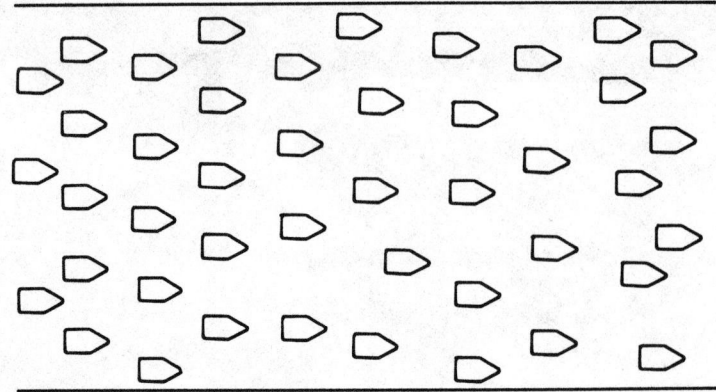

Figure 2-6 During recording, the magnetic particles line up with the magnetic field.

head, and therefore, its magnetic field are driven by an ac (fluctuating) signal—specifically, a varying voltage that represents an analog of the sound wave to be recorded. Consequently, the alignment of the magnetic particles on the tape varies along the length of the tape according to the varying voltage (and magnetic field) that corresponds directly to the instantaneous amplitude of the wave form being recorded. This is illustrated in Fig. 2-7.

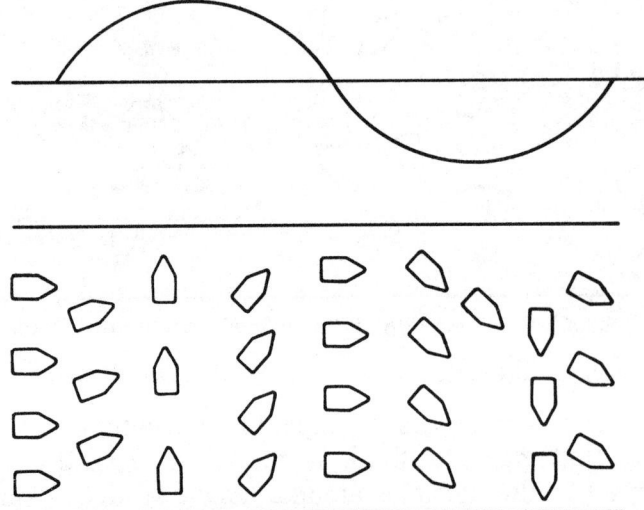

Figure 2-7 The position of the magnetic particles at any point along the length of the tape indicates the instantaneous amplitude of the recorded signal at that point.

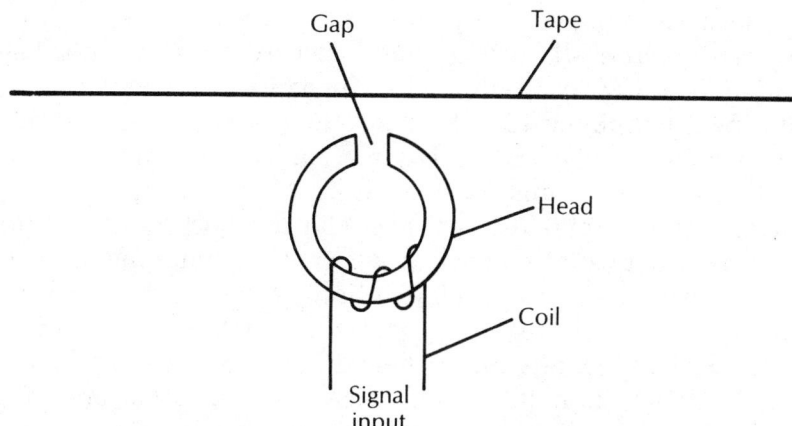

Figure 2-8 *Magnetic tape is recorded by passing a varying voltage through a coil.*

The fluctuating magnetic field is created by feeding the varying analog voltage to a device called a record head, which is basically just a specialized coil enclosed in a metallic housing. The signal voltage is fed through a coil, causing a varying magnetic field due to the induction effect. A simplified diagram of the way a typical record head is constructed is shown in Fig. 2-8.

The playback head

Now that the analog information representing the original sound waves has been magnetically stored on the tape in the form of patterns of alignment of the magnetic particles suspended in the tape's binder, what can be done with it? How is the recorded information converted back into sound waves for playback of the recording?

As it happens, the playback process is really very similar to the recording process, except, not surprisingly, it works in reverse. A varying voltage through a coil can create a fluctuating magnetic field. This is the operating principle behind the record head. The same principle works in the opposite direction too. A varying magnetic field placed across a coil will induce a proportionately fluctuating voltage in the coil. The playback head works in exactly the opposite way as the record head. The alignment (polarity) of the magnetic particles on the tape is sensed and converted into a varying analog voltage, which is then amplified and converted back into sound waves via a loudspeaker.

Because the operating principles are the same, a single device can theoretically act as either a record head or a playback head, depending on whether a varying voltage or a varying magnetic field is applied to it. In many low-priced tape recorders, the record and playback heads are, in fact, combined into a single, inexpensive dual-purpose unit. This is inexpensive and convenient, but not really very desirable for most serious recording work. There are times when we want to simultaneously use both the record and playback functions, such as when monitoring one track while recording a second track. A single record/playback head severely limits the editing capabilities of the tape recorder. In addition, the specific design requirements are different for record heads and playback heads. The gap size is one of the crucial factors.

The erase head

There is a third type of head that is included in all practical tape recorders. (Although it is always omitted in playback-only units.) This is the erase head. It is basically a simpler version of the record head, and is driven by a high-frequency (beyond the normal recording capabilities of the machine) signal called bias. This ensures that the magnetic particles on the tape are more or less randomly aligned before recording.

If a pre-existing alignment pattern (an earlier recording) is present, it might not be entirely canceled out and replaced by the alignment pattern for the newly recorded signal. On playback, you would then hear a muddy blend of the two recordings, which is obviously undesirable. The erase head makes sure the tape is a blank slate for the record head to write on.

For some reason, which is not fully understood by engineers, a small amount of the bias signal improves the recording quality. This is done on all modern tape recorders. Recording bias will be discussed in a little more detail in a later section of this chapter.

The arrangement of the heads

All tape recorders (except for playback-only units) have two or three heads. Inexpensive models have just two heads: an erase head and a record/playback head.

Note that the heads are always arranged in this specific or-

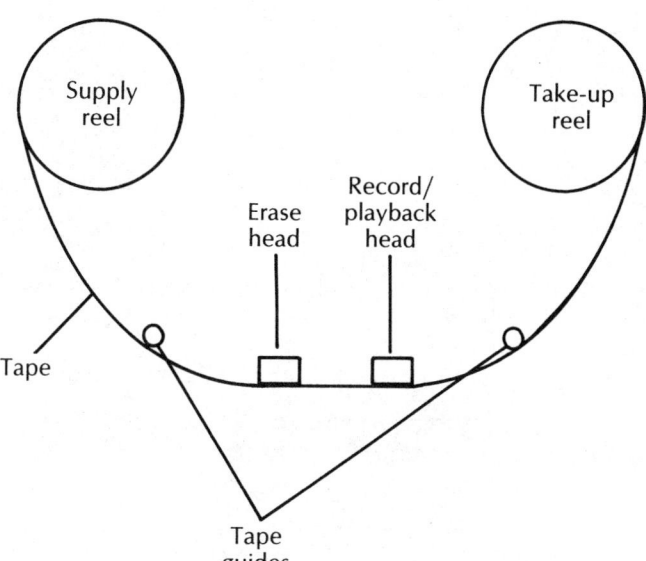

Figure 2-9 *Two-headed tape recorders have their heads in this order.*

der, as illustrated in Fig. 2-9. This really makes quite a bit of sense when you think about it. If the erase head was positioned after the record/playback head, any signal recorded on the tape would immediately be erased as it passed over the erase head. Obviously this wouldn't be a particularly useful piece of audio equipment.

As indicated earlier, better quality tape recorders designed for more serious use generally have three heads: an erase head, a record head, and a playback head.

Once again, the three heads are always arranged in this exact order, as illustrated in Fig. 2-10. The erase head must come before the record head to prevent immediate erasure of any signal as soon as it is recorded. The playback head is positioned immediately after the record head to permit the user to monitor the recorded signal an instant after it has been recorded—an extremely useful feature. By monitoring the signal as it is recorded, any possible problems can be quickly spotted and corrected.

Factors determining frequency response

In any sound reproduction system, frequency response is a very important consideration. Ideally, all frequencies within the audible range should be treated equally. That is, no frequency, or

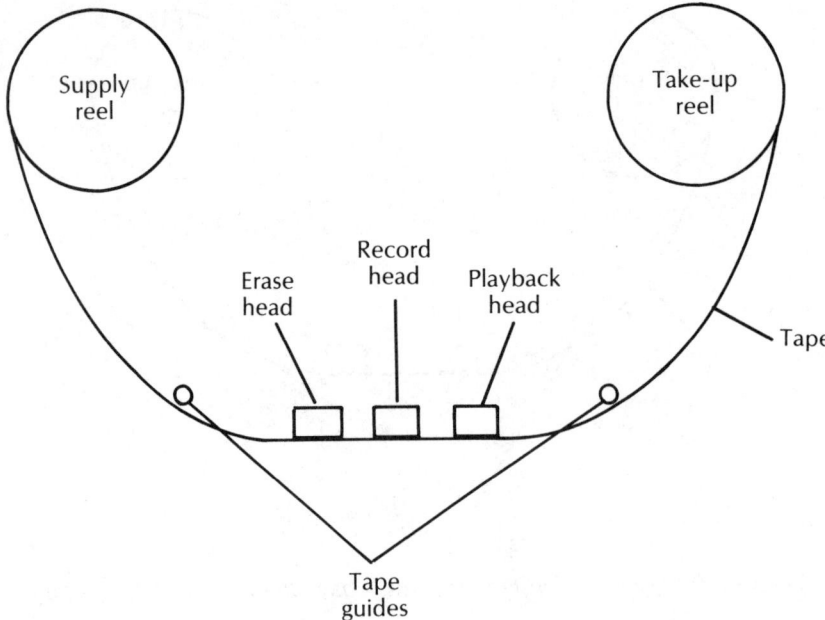

Figure 2-10 *All three-headed tape recorders position their heads in this order.*

band of frequencies, should recieve any more amplification or attenuation than any other frequency. This is vitally important if the playback signal is to properly duplicate the original signal that was recorded. Consider for a moment how badly music sounds coming through a cheap AM pocket radio. This is because of the extremely limited frequency response. A cheap portable AM radio is more or less the worst of all worlds when it comes to limited frequency response; the limitations come in every step of the way. To start with, AM broadcast signals are severely frequency limited, according to FCC regulations. This is done to fit the maximum number of stations into the available bandwidth. Then, since this is a cheap radio, both the receiver section and the audio amplifier are probably "quick and dirty" designs—economy, not quality, was the goal. Each stage of the circuitry undoubtably eats away at the overall frequency response of the signal. Finally, a very small and cheap speaker is used to reproduce the sound. It's just not up to the job. It can't reproduce very low frequencies or very high frequencies. Even the frequencies it can reproduce it undoubtably reproduces very unevenly. The frequency response of such a tiny speaker is surely far from flat. This is an extreme example of the importance of frequency response in audio reproduction.

Graphing the frequency response of an ideal tape recorder would result in a straight line, as shown in Fig. 2-11. Unfortunately, this ideal frequency response is very easy to describe and graph, but quite impossible to achieve in any practical analog system.

Very low frequencies tend to be attenuated because the fluctuations are too slow and gradual to be decoded properly. Very low-frequency signals run the risk of looking like random noise to the tape recorder's heads.

Even more crucial is the high-frequency roll off. High frequencies tend to become sharply attenuated because the signal fluctuations are too brief. Not enough tape passes by the record head to record the cycle pattern. The magnetic particles again appear to be randomly aligned, rather than aligned to a repeating pattern. High-frequency signals tend to get lost in the random white noise hiss inherent in all analog tape recordings.

The frequency response graph for a practical tape recorder is shown in Fig. 2-12. The maximum high frequency that can be recorded and played back by a given tape recorder is dependent on many different factors within the system. One factor is the gap size of the recording head. Everything else being equal, narrower gaps can record higher frequencies. Another very important (and usually more user-controllable) factor in defining the

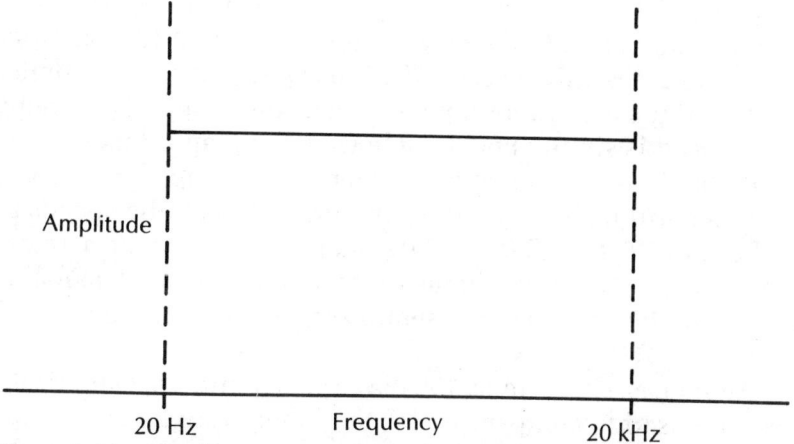

Figure 2-11 *The frequency response of an ideal tape recorder would be graphed as a straight line.*

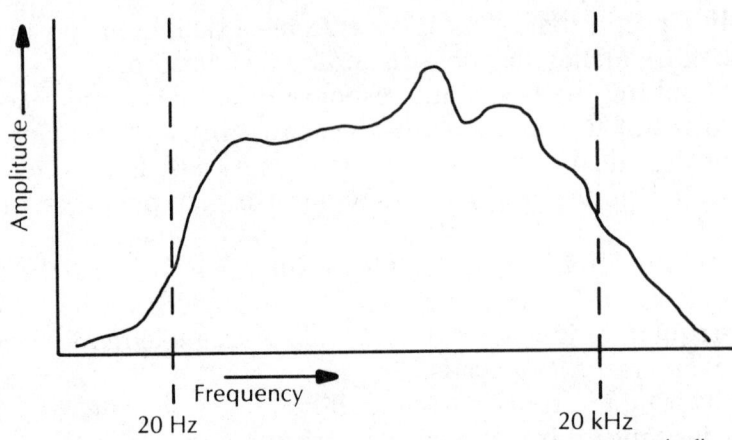

Figure 2-12 *A practical tape recorder does not have a truly flat frequency response.*

high-frequency response of a given tape recorder is the speed of the tape past the heads. The faster the tape moves past the head, the more space there will be to encode the wave form.

Modern home reel-to-reel recorders usually have two speeds: 3¾ IPS and 7½ IPS. The term IPS, you should recall, means inches per second.

The slower speed is used when tape economy is the primary consideration. A recording of a given length takes twice as much tape at 7½ IPS as at 3¾ IPS. But the higher speed is generally preferred for high-fidelity recording, because of the improved high-frequency response.

Some reel-to-reel recorders also include a still slower speed: 1⅞ IPS. The use of this speed should be limited to noncritical voice recordings only. The 1⅞ IPS tape speed is very economical in terms of the amount of tape used, but the frequency response at this speed is simply not suitable for music reproduction.

Professional studio tape recorders usually operate at higher speeds for still higher fidelity. Commonly used studio speeds are 15 IPS and 30 IPS. Semiprofessional recorders offer a sort of compromise between the home recorder and the professional recorder. The two speeds on a semiprofessional unit are typically 7½ IPS and 15 IPS.

Audio cassettes run at 1⅞ IPS. Special circuitry and design tricks are used to compensate for the inherent frequency limitations of the slow tape sped. Eight-track cartridges are recorded at 3¾ IPS.

The amount of tape needed to hold one cycle can be calculated with this simple formula:

$$L = S/F \qquad (1)$$

Where:
   L = wavelength space on the tape, in inches;
   S = tape speed relative to the record head, also in inches; and
   F = frequency of the signal to be recorded (in Hz).

Clearly, the greater the space used to record a cycle, the more detail there can be in representing the wave shape.

To better understand how this works, let's consider some practical examples. We will assume that we need to record a top frequency of 12 kHz (12,000 Hz). At 1⅞ (1.875) IPS, the amount of tape to record 1 cycle of a 12-kHz signal works out to

$$L = 1.875/12{,}000$$
$$= 0.000156 \text{ inch}$$
$$= 0.156 \text{ milliinches}$$

That certainly isn't very much space to define the wave shape.

Increasing the recording speed to 3¾ (3.75) IPS increases the available 1 cycle space to

$$L = 3.75/12{,}000$$
$$= 0.000312 \text{ inch}$$
$$= 0.312 \text{ milliinches}$$

Then, at 7½ (7.5) IPS

$$L = 7.5/12{,}000$$
$$= 0.000625 \text{ inch}$$
$$= 0.625 \text{ milliinches}$$

At 15 IPS

$$L = 15/12{,}000$$
$$= 0.00125 \text{ inch}$$
$$= 1.25 \text{ milliinches}$$

And, finally, at 30 IPS

$$L = 30/12{,}000$$
$$= 0.0025 \text{ inch}$$
$$= 2.5 \text{ milliinches}$$

This is unquestionably a very considerable improvement over the 0.000156-inch space allowed by a speed of 1⅞ IPS. Higher tape speeds allow much more room for recorded detail (better fidelity), especially for higher-frequency signals.

Voltage signal

Playback in a tape recorder is accomplished by magnetically inducing a voltage into the playback head. This induced voltage is extremely small, and must be considerably amplified to be a useful signal. Depending on the specifics of the design, the required amplification for this signal may be from 10 to 10,000. Narrow track widths and small head gaps generally call for greater amounts of amplification. Both a narrow track width and a small head gap leave less room for the signal, so it follows that the resulting signal must be weaker than if there was more room available.

The weakness of the induced voltage signal in the playback head aggravates the ever-present noise problem. In any electronic circuit, some random noise signal is inevitably and inescapably generated. This noise signal is typically rather small and is often totally negligible in many practical applications. But in cases where the desired signal is also very small (as in the case of the induced voltage from a tape recorder's playback head), the noise signal can become quite significant.

The problem is compounded because there is also some noise signal being picked up from the tape by the playback head itself. Not all of the magnetic particles on the tape will be properly aligned with the recorded (desired) signal. Some will always be more or less randomly oriented. This is especially true when the recorded signal has a relatively low amplitude.

At the opposite extreme, if you attempt to record a signal at too high an amplitude, the tape will be overloaded beyond its storage capacity, and distortion will result on playing the tape back.

To make matters even worse, the induced voltage in the playback head is at its weakest for very low and very high frequencies. Midrange frequencies are picked up the best. At extreme frequencies (both high or low), noise tends to be at its strongest levels.

Typically, loudspeakers also exhibit their worst frequency

response at the extreme ends of the audible frequency spectrum (very low frequencies and very high frequencies). It is also worthwhile to remember that the human ear—even if it is working perfectly—is not a very linear device. It's own frequency response is decidedly nonflat. The human ear tends to be most sensitive to midrange frequencies. Low-frequency and high-frequency sounds need to have higher amplitudes to be heard at the same perceived volume as a weaker midfrequency sound.

Very careful design throughout all stages in a tape recorder is an absolute must to reduce the inherent noise signals down to a tolerable level. Most modern tape recorders do quite well in this respect, at least once you get out of the budget category. But, of course, some designs do considerably better than others.

The amount of magnetic charge, or flux, on the tape varies with the recorded frequency. Higher-frequency signals generally tend to result in increasingly lower flux, as illustrated in the graph of Fig. 2-13.

The voltage signal induced into the recorder's playback head is dependent primarily on two factors: the amount of tape flux and the frequency of the recorded signal. The playback head is designed so that its frequency response partially compensates

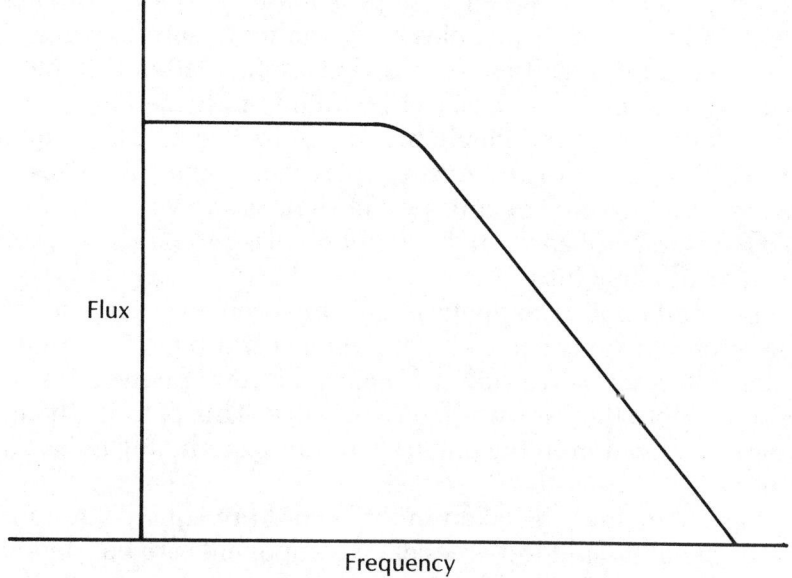

Figure 2-13 *The higher the frequency, the lower the flux.*

68   Analog Sound Recording

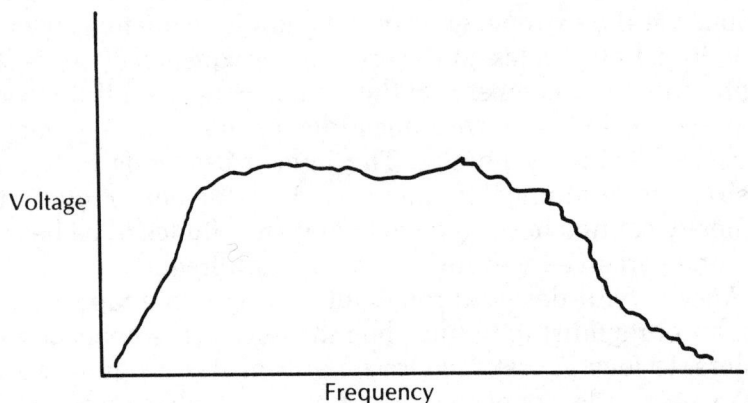

Figure 2-14   *This is a voltage versus frequency graph for a typical tape recorder.*

for the decreasing flux at higher frequencies. A typical voltage versus frequency graph is shown in Fig. 2-14.

Equalization

To further improve the overall performance of a tape recorder system, and to come closer to the ideally flat frequency response, some sort of equalization is usually applied to the signal. *Equalization* is a deliberate unequal amplification of different frequencies. Low and high frequencies are given more amplification, or boost, than midrange frequencies. This helps flatten out the recorder's frequency response, as illustrated in Fig. 2-15.

Unfortunately, and inevitably, any noise signal is also equalized right along with the desired playback signal. As a result, low- and high-frequency noise is boosted, as shown in Fig. 2-16. This tends to emphasize such common noise problems as amplifier hum and tape hiss.

In an attempt to compensate for this problem, most modern tape recorders apply some sort of pre-equalization to the original signal before it is recorded. High and low frequencies are boosted before they are applied to the tape. This permits the use of less equalization in the playback circuitry, reducing noise emphasis.

Great care must be taken in the record pre-equalization process. If the high- and low-frequency components are boosted too much, the tape and the electronic circuitry may be overloaded, resulting in distortion of the recorded signal.

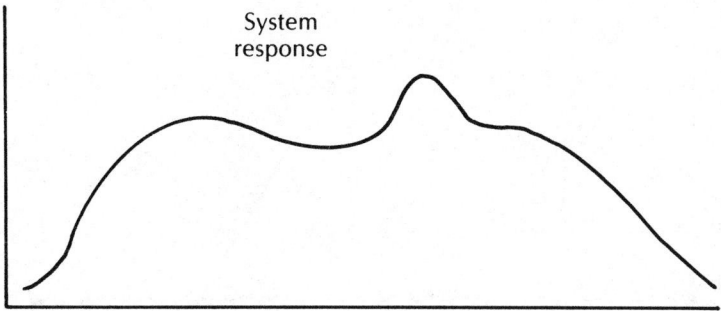

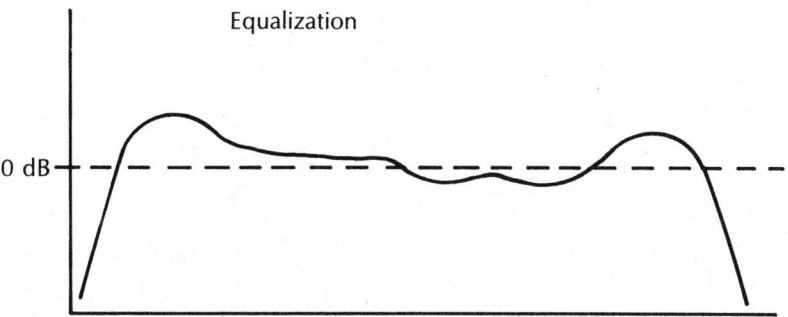

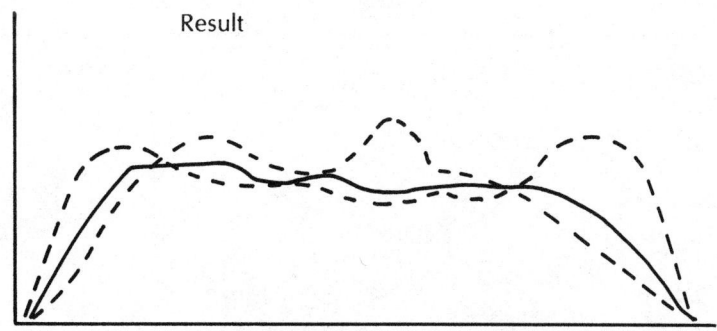

Figure 2-15 *Boosting the lowest frequencies and the highest frequencies can help flatten out the frequency response.*

The audio industry has set up equalization standards based on what is known about sound level versus frequency in speech and music. These equalization techniques help flatten out the system's overall frequency response. They also help to improve the signal:noise ratio.

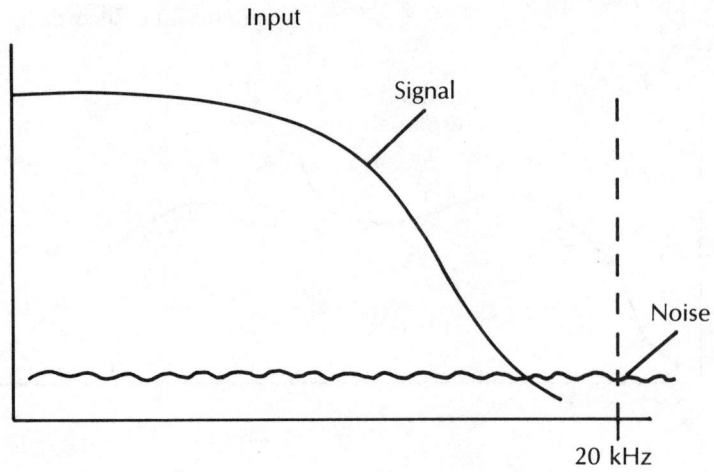
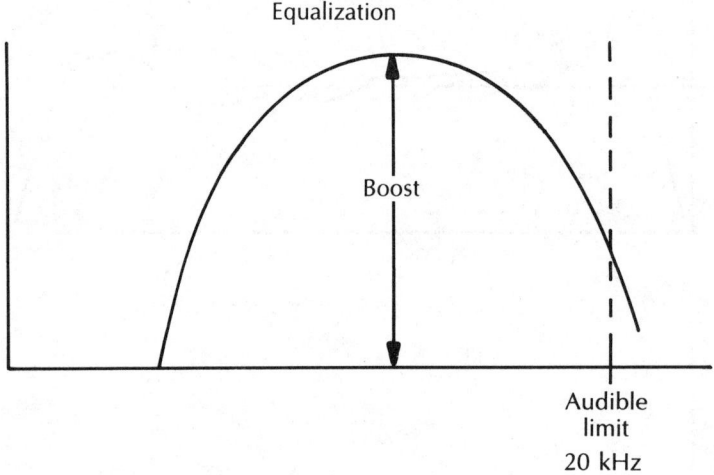
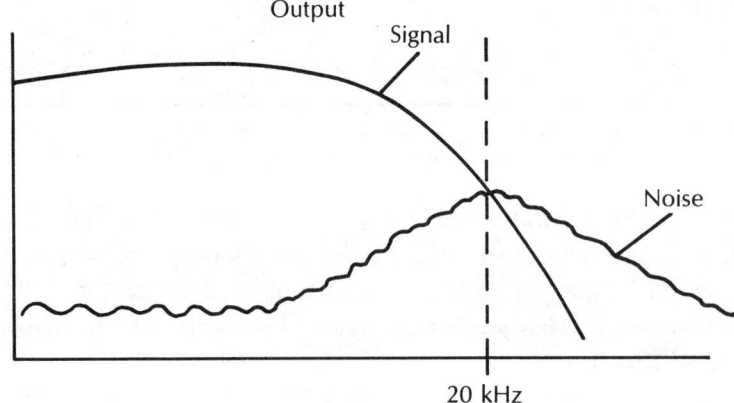

Figure 2-16 *Noise is usually boosted along with the desired signal.*

The equalization standard used in most consumer tape recorders is the NAB (National Association of Broadcasters) equalization standard. The NAB equalization standard is similar in concept to the RIAA equalization standard used for vinyl records. The specific frequency points and the boost and cut slopes are somewhat different for the two standards.

The signal:noise ratio is a very important specification for any type of audio equipment. The stronger the desired signal is with respect to the noise level, the better the equipment is. The signal:noise ratio is generally given in decibels (dB). The reference is the noise level. For example, for a tape recorder with a signal:noise ratio of 44 dB, the recorded (desired) signal is 44 dB higher in amplitude than the noise signal. If all other factors are equal, the signal:noise ratio is degraded at least 3 dB if the track width is halved (for example, four tracks as opposed to two for tape of a given width).

While there is considerable variation due to several design factors, the signal:noise ratio for typical sound reproduction systems are as follows:

- 78-rpm phonograph records   35–40 dB
- Inexpensive tape recorders   35–45 dB
- Good-quality home tape recorders   45–55 dB
- Modern LP (33⅓ RPM) phonograph records   50–65 dB
- Studio tape recorders (analog)   50–75 dB
- Digital studio recorders   85–95 dB

A good equalization system can give a tape recorder a frequency response very close to flat. However, a truly flat frequency response along the entire audible spectrum is just not possible. There are simply too many factors involved.

The frequency response specification for a tape recorder is typically given in this form:

$$X \text{ to } Y \text{ Hz} \pm Z \text{ dB} \qquad (2)$$

Where:
 X = lowest measured frequency,
 Y = highest measured frequency, and
 Z = maximum variation from flat response in the specified frequency range.

For example, a fairly typical tape recorder might have a specified frequency response of 60 to 15,000 Hz ± 3 dB. This means that any frequency between 60 and 15,000 Hz (15 kHz) will be no more 3 dB higher or 3 dB lower in amplitude than the nominal flat frequency response level. A difference of 3 dB is just barely noticeable for most people.

In shopping for a tape recorder, or any other audio equipment, look for the widest possible specified frequency range, with the lowest decibel variation figure. As a rule of thumb, consider the following frequency response decibel values:

- 1 dB—Excellent frequency response; just barely noticeable by an expert under ideal listening conditions.
- 3 dB—Good frequency response; noticeable to most people under normal listening conditions.
- 6 dB—Poor frequency response; an unmistakable change in sound level at different frequencies under normal listening conditions.

Frequencies outside the specified range can usually be reproduced by the tape recorder, but the variation in level may be greater than specified. For instance, in our sample machine with a frequency response of 60 to 15,000 Hz ± 3 dB, a 17,500-Hz signal might be reproduced at a level that is 5 dB lower than the nominal flat response level.

Few home tape recorders are specified for use above about 15,000 to 16,000 Hz (15 to 16 kHz). This might seem to be a major limitation, since the human ear nominally can hear frequencies up to 20,000 Hz. Fortunately, this is not nearly as big a problem as it might seem at first. The maximum audible frequency tends to decrease with the listener's age, and few serious listeners can really hear much above 15 to 16 kHz. Besides, few musical sounds contain strong frequency components above about 10,000 Hz. A trained listener or audiophile will probably be able to perceive some difference, but for most general listening, a frequency response of about 60 to 15,000 Hz is usually considered entirely acceptable and even pretty good for analog sound reproduction.

Record bias

An extremely important factor in reproducing sound from an analog tape recorder is the record bias. This is an ultrasonic (inaudible) high-frequency signal generated by an internal oscillator. This ultrasonic bias signal is mixed with the audible signal to be recorded at the record head. Without getting into the theoretical aspects, which are somewhat complicated, the record bias increases the recorder's overall high-frequency response and decreases distortion in the recording.

The exact frequency of the record bias signal is not particularly critical. It must only be high enough not to cause audible beating effects. Beating occurs at the difference between two simultaneous frequencies. For example, say you are recording a 12-kHz signal, and the record bias is set at 17 kHz. A spurious beat frequency of 5 kHz might appear in the recorded signal. Clearly, this is extremely undesirable.

In most modern analog tape recorders, the record bias frequency is set at 70 to 150 kHz. This range of frequencies does the job quite well and without any obtrusive side effects. Often the same high-frequency signal that is used by the erase head to erase a previously recorded signal is also employed as the record bias signal. Of course, the bias level is much lower at the record head than at the erase head. Obviously, there wouldn't be much point in simultaneously recording and erasing the signal.

While the record bias frequency is not critical, the amplitude of the record bias signal is. If the record bias level is set too low, or if there is no bias used at all, the recorded signal will be distorted. If you're lucky, this distortion may be fairly mild, resulting in nothing worse than a rather bland sound on playback. But often the distortion may be quite severe and very objectionable. In addition, the signal-to-noise ratio will be significantly degraded if the bias signal is not strong enough.

On the other hand, if the bias signal level is set too high, the record head will tend to function more as an erase head than a record head. This is especially likely when the signal to be recorded includes mostly high audible frequencies. The signal will be erased as it is being recorded.

Unfortunately, determining the best record bias level is not an easy task. Many factors must be considered. These include, but are not limited to

- Design of the record head,
- Design of the recorder's electronic circuitry,
- Characteristics of the tape being used, and
- Frequency of the signal being recorded.

The first two factors remain fairly constant, although some of these characteristics might change as the components age. At any rate, the designer of the recorder has the most control here.

Tape characteristics, however, can differ significantly from brand to brand, or even from batch to batch for the same brand. The frequency of the signal being recorded will also vary a great deal. The designer of a tape recorder obviously has no control at all here.

Because of these problems, the theoretical "ideal" record bias level is doomed to eternally remain an "impossible dream." Fortunately, a good, reasonable compromise can usually be obtained without too much difficulty. As a rule of thumb, the record bias signal's amplitude is set at a level approximately 10 times the amplitude of the signal to be recorded.

Because of record bias variations, any debate over the best recording tape is almost always a moot point. The best tape for tape recorder A probably is not the best tape for tape recorder B.

WOW AND FLUTTER

Analog recording systems are very sensitive to speed fluctuations. This goes for tape recorders of any format, as well as for turntables.

The speed at which a turntable rotates or a tape is pulled past a recorder's heads is critical, and must be precisely maintained. This is generally easier said than done. Both turntables and tape recorders are prone to minor speed fluctuations.

If the speed fluctuates at a fairly low (subaudible) frequency, the effect is called wow. If the speed fluctuates at a higher (audible) frequency, the effect is known as flutter. Both of these terms are fully self-explanatory once you've heard them in action. A bad case of wow will give the sound a "wow-wow-wow" effect. When the problem is flutter, the sound has a distinct "fluttering" quality.

Obviously, the lower the wow and flutter ratings for a turntable, or a tape recorder, the better the equipment is. Wow and flutter tend to be more of a problem for cassette and 8-track recorders than for reel-to-reel machines. This is because the enclosed housing (cassette or cartridge) can contribute to the problem if the parts aren't perfectly fitted together. The endless loop configuration within an 8-track cartridge seemed particularly prone to severe wow problems.

❖ 3
Basics of Digital Recording

DIGITAL AUDIO RECORDING IS INHERENTLY MORE COMPLEX THAN analog audio recording. The reason for this is that sound itself is an analog (or linear) phenomena. Changes in amplitude can be graphed along a continuous straight line, as shown in Fig. 3-1. An intermediate value can always be inserted between any two adjacent values. For example, between 2 and 3, there is 2.5, between 2.5 and 3 is 2.75. Between 2.75 and 3 is 2.875, and so on, into infinitely small increments.

In a digital system, however, only whole numbers can be used. Increments follow a decidedly steplike pattern, as illustrated in Fig. 3-2. There is no such thing as an intermediate value between, say, 2 and 3. It has to be either 2 or 3—in-between values are impossible.

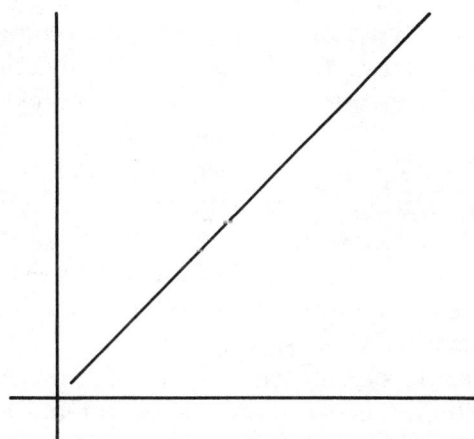

Figure 3-1 *An analog signal can be graphed as a straight line.*

76   Basics of Digital Recording

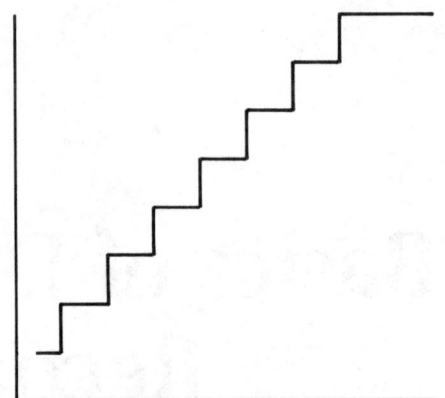

Figure 3-2   *Digital signals are in the form of discrete steps.*

All this probably seems very awkward, and digital audio recording may seem like a totally pointless exercise at first glance. However, as we shall see, digital recording offers some very important advantages over the simpler analog recording techniques.

In this chapter we will explore the basics of digital audio recording in general. Chapter 4 will describe how these principles are used with compact discs(CDs), and Chapter 5 will discuss the specific details of the DAT system.

NUMBERING SYSTEMS

Any value can be used as the base of a numbering system. It may sound like we're getting into some complicated mathematics here, but we're really not. If you can count from one to ten, you should be able to grasp the fundamental concepts introduced here. Admittedly, converting from one numbering system to another can be fairly difficult. Fortunately, you're not going to have to do any of that. You do not need to be able to perform any mathmatical calculations to understand or use digital audio. Our goal here is simply to build an understanding of the basic principles involved. The word *digital* implies the use of numbers, so any digital process (including digital audio recording) is, by definition, a mathmatical process of some kind.

The numbering system we are all used to working with is the decimal system. It is based on a value of ten. *Decimal* simply means "based on ten."

We are so used to the decimal numbering system that most

people aren't even aware other numbering systems are possible, and often these alternatives may be more useful and practical than the more familiar and comfortable decimal numbering system.

The decimal numbering system is very convenient for human beings to use, because most human beings happen to have ten fingers. We can count in decimal on our fingers. There is no more significance to use of ten as the base value than that. There is nothing magical or inherently special about the value ten. It just seems "natural" because of long established cultural habits. Any other base value would work every bit as well.

In the decimal numbering system, there are ten available digits, ranging from zero to nine (0–9). All decimal numbers must be made up of some combination of these ten digits.

What if we need to represent a value larger than nine? Simple. We just add a new column to the left. This new column is in multiples of the base (10) rather than units. For example,

$$54 = (5 \times 10) + (4 \times 1)$$

If we go past a value of 99, we have to add a third column to the number. The value of each new added column is equal to the value of the preceeding column multiplied by the base (ten). So, in the decimal numbering system, the third column is equal to 10×10, or 100; the fourth column is equal to 100×10, or 1000; the fifth column is equal to 1000×10, or 10,000; and so forth. This can be extended as far as necessary to express the needed number. As an example,

$$\begin{aligned}41{,}736 &= (4 \times 10^4) + (1 \times 10^3) + (7 \times 10^2) + (3 \times 10) + (6 \times 1) \\ &= (4 \times 10 \times 10 \times 10 \times 10) + (1 \times 10 \times 10 \times 10) \\ &\quad + (7 \times 10 \times 10) + (3 \times 10) + (6 \times 1)\end{aligned}$$

Now, it may well seem that we are making things unnecessarily complicated. All of this just comes naturally to most of us when we're working with decimal numbers. We know what 41,736 means when we see it. Why bother breaking the number down like this? It seems like a lot of fuss and bother for nothing.

Actually, we do the same kind of thing unconsciously whenever we recognize a multidigit number. Consciously understanding how to break a multidigit number down this way becomes very helpful when we move on to alternate numbering systems.

Earlier we said that any base value can be used for a numbering system. In digital electronics, the binary numbering system is normally used. *Binary* simply means "based on two." In the binary numbering system there are just two digits—0 and 1.

The binary numbering system was not selected because electronics engineers are perverse and wanted to make things complicated and hard to understand. As it happens, the binary numbering system is a very easy one for electronic circuits to work with. A 0 can be represented by the absence of a voltage, and a 1 can be represented by the presence of a voltage. The circuitry doesn't have to bother with any awkward (and possibly difficult to measure) intermediate voltages.

In effect, digital electronic circuitry has just "two fingers," so the binary numbering system makes more sense to it than the decimal numbering system. Binary values can also be thought of as yes or no indicators. A 1 means yes, and a 0 means no. Ultimately, the numbers can mean anything we want them to mean.

While it may not appear obvious, the binary numbering system works just like the decimal numbering system. When you need to express a number larger than the largest available digit (1 in this case), you have to add more columns to the left. For example,

$$\begin{aligned}
1101101_2 &= (1 \times 2^6) + (1 \times 2^5) + (0 \times 2^4) + (1 \times 2^3) \\
&\quad + (1 \times 2^2) + (0 \times 2) + (1 \times 1) \\
&= (1 \times 2 \times 2 \times 2 \times 2 \times 2 \times 2) + (1 \times 2 \times 2 \\
&\quad \times 2 \times 2 \times 2) + (0 \times 2 \times 2 \times 2 \times 2) \\
&\quad + (1 \times 2 \times 2 \times 2) + (1 \times 2 \times 2) + (0 \times 2) \\
&\quad + (1 \times 1) \\
&= (1 \times 64) + (1 \times 32) + (0 \times 16) + (1 \times 8) \\
&\quad + (1 \times 4) + (0 \times 2) + (1 \times 1) \\
&= 64 + 32 + 0 + 8 + 4 + 0 + 1 \\
&= 109_{10}
\end{aligned}$$

The subscripts (the small 2 and 10 to the lower right of the expressed numbers) indicate the base of the particular numbering system being used. That is 1101101_2 in binary equals 109_{10} in decimal.

Table 3-1 compares some binary and decimal numbers. Notice that leading zeroes are commonly used in the binary numbering system, but are omitted in the decimal numbering system.

Table 3-1 Comparison of binary and decimal numbers.

Binary	Decimal
0000	0
0001	1
0010	2
0011	3
0100	4
0101	5
0110	6
0111	7
1000	8
1001	9
1010	10
1011	11
1100	12
1101	13
1110	14
1111	15
0001 0000	16
0001 0001	17

This custom simply reflects the way digital electronic circuitry works. Any digital circuit has a fixed number of digits. If the circuitry is set up for four bits, all four places must be filled with either a 0 or a 1. No space can be left blank or omitted. The use of leading zeroes also helps make binary numbers stand out visually as something other than common decimal numbers.

Each digit in a binary number is called a bit, or BInary digIT. Eight bits makes up a byte. Four bits form a nybble.

If you read a lot about digital electronics, you'll probably occasionally come across references to the octal numbering system and the hexadecimal numbering system. These are compromise numbering systems used to make the binary numbering system more convenient for human beings to work with.

If we group bits into sets of three, each unit can take on any of seven possible values, from 000 to 111. This gives us the octal (base 8) numbering system, with eight possible digits—zero through seven (0–7). Notice that there is no 8 or 9 in the octal numbering system. Table 3-2 compares the binary, octal, and decimal numbering systems.

The hexadecimal (base 16) numbering system is quite similar to the octal numbering system, but in this case the binary bits are grouped into sets of four instead of three. This gives sixteen

Table 3-2 Comparison of binary, octal, and decimal numbering systems.

Binary	Octal	Decimal
000 000	0	0
000 001	1	1
000 010	2	2
000 011	3	3
000 100	4	4
000 101	5	5
000 110	6	6
000 111	7	7
001 000	10	8
001 001	11	9
001 010	12	10
001 011	13	11
001 100	14	12
001 101	15	13
001 110	16	14
001 111	17	15
010 000	20	16
010 001	21	17

possible unit values from 0000 to 1111. New digits have to be used for values from 10 to 15, because such single digits do not exist in the common decimal numbering system. The letters A through F are used for this purpose—zero through nine (0–9) and A through F (A–F).

Table 3-3 compares the binary, hexadecimal, and decimal numbering systems.

Unless you are working with the actual design of digital circuits or writing computer programs, you will probably never have to convert between one numbering system and another. However, a rough comprehension of the basic concepts involved is a big help in understanding how a digital system, such as digital audio recording, works.

Just remember, when we write down a number, we are merely using standardized symbols to represent a value. The symbol is not what it represents. The same symbol can be used to mean something else under different circumstances.

ANALOG SOUND INTO DIGITAL BITS

In digital audio recording, the instantaneous amplitude (level) of the analog sound is repeatedly sampled (thousands of times per

Table 3-3 Comparison of the binary, hexadecimal, and decimal numbering systems.

Binary	Hexadecimal	Decimal
0000 0000	0	0
0000 0001	1	1
0000 0010	2	2
0000 0011	3	3
0000 0100	4	4
0000 0101	5	5
0000 0110	6	6
0000 0111	7	7
0000 1000	8	8
0000 1001	9	9
0000 1010	A	10
0000 1011	B	11
0000 1100	C	12
0000 1101	D	13
0000 1110	E	14
0000 1111	F	15
0001 0000	10	16
0001 0001	11	17
0001 0010	12	18
0001 0011	13	19
0001 0100	14	20
0001 0101	15	21
0001 0110	16	22
0001 0111	17	23
0001 1000	18	24
0001 1001	19	25
0001 1010	1A	26
0001 1011	1B	27
0001 1100	1C	28
0001 1101	1D	29
0001 1110	1E	30
0001 1111	1F	31
0010 0000	20	32
0010 0001	21	33
0010 0010	22	34
0010 0011	23	35

second). The measured amplitude at that instant is converted into the nearest digital value. In other words, the continuous wave form is converted into a discontinuous series of discrete numbers, as illustrated in Fig. 3-3.

This string of binary numbers representing the recorded wave form can then be electronically stored (and possibly manipulated) by digital circuitry. This is a very simplistic description of digital audio recording.

82 Basics of Digital Recording

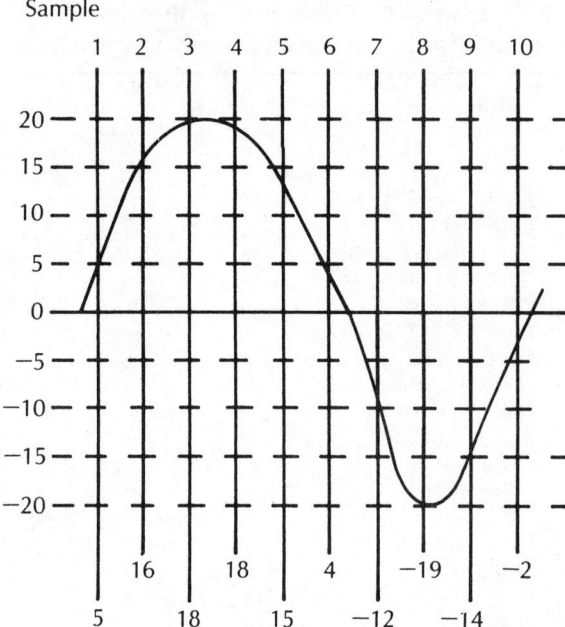

Figure 3-3 A continuous wave form (analog) is converted into a discontinuous series of discrete numbers (digital).

Playback works in just the opposite way. Each number in the sequence is converted into a proportionate analog voltage. Many numbers in rapid succession can simulate an analog wave form, as illustrated in Fig. 3-4. A filter can be used to smooth off the sharp edges of the digital "steps," as illustrated in Fig. 3-5.

DIGITAL RECORDING SIGNALS

Once the analog signal has been converted into digital form, another problem arises—just how do we store (record) all those 1s and 0s? If we are just using an electronic (computer) memory, there's no problem storing data in digital form. In fact, in that case, we have no choice. All data must be in digital form.

Musical recordings take up a lot of memory. If that's all we use, the system would be incredibly expensive and unwieldy, even if we tried to store just one or two three minute songs. Actually, such direct electronic storage is used only for short-term recording of limited duration (a few seconds), mainly for manipulation of the stored sound.

A specialized media, such as the CD, can sometimes be de-

Digital Recording Signals 83

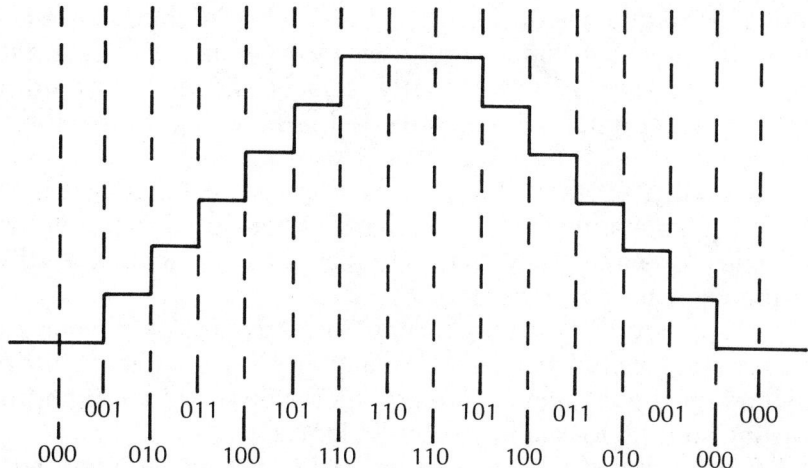

Figure 3-4 *Many numbers in rapid succession can simulate an analog wave form.*

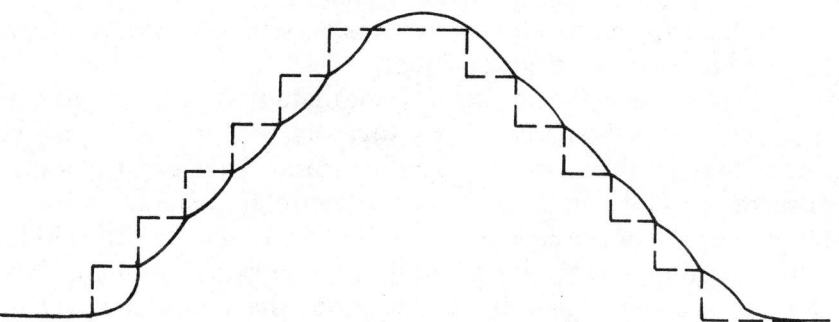

Figure 3-5 *A filter can "smooth out" the sharp corners of discrete digital steps.*

signed to store digital data (1s and 0s) directly. For example, on a CD, a pit indicates a 0 and an island indicates a 1. (This is a very simplified description. Actually, digital data on a CD is modulated and encoded in a fairly complex manner. This will be discussed in more detail in Chapter 4.)

Tape recording is very handy, efficient, and relatively inexpensive. It can be easily recorded and played back. Old, no longer needed recordings can be conveniently erased and new material can be recorded onto the tape. Unfortunately, a 1 or a 0 just can't be recorded onto a strip of magnetic tape. Some sort of modulation is needed.

One of the simplest approaches to recording digital data is

known as frequency shift keying (FSK). This method is used by many inexpensive computers to store programs and data onto ordinary cassette tapes. Actually, this is an analog recording method. The digital data is converted into analog form before it is recorded.

Basically, in frequency shift keying, an analog signal is recorded. This analog signal may have either of two possible frequencies. One frequency is used to record a 0 bit, while the other frequency is used to record a 1 bit.

Frequency shift keying is not normally used for recording music. As I stated earlier, it is actually an analog recording method for digital signals. We are interested here in a digital recording method for analog (sound) signals.

In practical digital recording systems, true modulation is used. Modulation, you should recall, involves a nominally constant wave form (the carrier signal), usually at a frequency well above the audible range. In the modulation process, some parameter of the carrier signal is varied in step with the audible signal to be recorded (the program signal).

To play back the recorded (modulated) signal, we use demodulation. As the name clearly suggests, this process is just the opposite of modulation. The original standardized carrier signal is recreated and compared with the modulated signal. Any difference between the recreated carrier signal and the full modulated signal makes up the original program signal. In demodulation, the carrier signal is deleted from the modulated signal, leaving the program signal.

In digital recording systems, the carrier signal is always in the form of a pulse wave. This may be called a rectangle wave or a square wave in some technical literature. Strictly speaking, there are some technical differences between these terms. Rectangle wave can refer to any two-stage signal that switches back and forth between a high level and a low level. In a square wave, exactly half of each cycle is high and half is low. The signal is in each of its two states for equal periods. A pulse wave generally is high for only a brief portion of each cycle. For consistency, we will use the term pulse wave in this book.

A pulse wave switches between two discrete (analog) voltages, as illustrated in Fig. 3-6. The transition time between the two states is extremely short and is considered negligible. In effect, the pulse wave switches instantly between states. A pulse

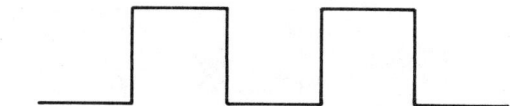

Figure 3-6 *A pulse wave switches back and forth between two discrete levels.*

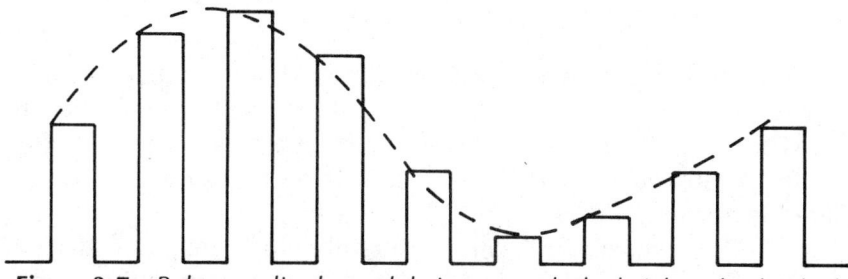

Figure 3-7 *Pulse amplitude modulation controls the heights of individual pulses.*

wave is used because it can readily be recognized and treated by either analog or digital circuitry.

There are many different ways the pulse wave carrier signal can be modulated by the program signal. If the height of the pulses are controlled (as shown in Fig. 3-7), we have pulse amplitude modulation (PAM). As you may have guessed from the name, this type of digital modulation is more or less the same as ordinary analog amplitude modulation (AM) (see Chapter 2). The only major difference between analog AM and PAM is in the carrier signal itself. Ordinary analog AM generally uses a sine wave. In PAM (as in all digital modulation systems), the carrier must be a pulse wave.

While PAM is conceptually similar to analog AM, the actual circuitry required to put the theory into practice is somewhat different.

Another approach is to modulate the percentage of each cycle that is in the high-voltage state, as illustrated in Fig. 3-8. Notice that the total cycle time (the signal frequency) remains constant. What is affected here is how each cycle is broken up into its high-voltage and low-voltage states. This system is called pulse width modulation (PWM).

Pulse width modulation is fairly popular in low- to middle-level communication systems, because this type of modulation is relatively easy to implement. Less commonly used systems

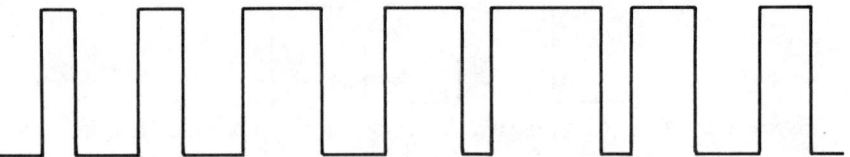
Figure 3-8 Pulse width modulation controls the duty cycles of individual pulses.

include pulse position modulation (PPM) and pulse number modulation (PNM). These names are fairly self-explanatory. In PPM the varying factor is the starting time of each pulse. How many pulses occur within a specific time period is the modulated factor in PNM.

PAM, PPM, and PWM are comparable to analog modulation, except that the carrier signal is a digitally compatible pulse wave. In PPN, however, the program signal is digitally encoded. The shape of the carrier pulses is not modulated.

The modulation method used in most digital sound recorders is pulse code modulation (PCM). Each sampled value from the original analog signal is converted into a pulse chain of a specific length. The analog signal, you should recall, is sampled thousands of times per second. A brief gap is left between pulse chains representing consecutive samples.

In PCM, as in PPN, the program signal is digitally encoded. The shape of the carrier pulses is not modulated. The PCM circuitry looks at each new sample of the digitized program signal and converts the sample value into a pulse chain of a specific length. In demodulation, these pulse chains are converted back into the appropriate sample values. PCM signals can be recorded with less bandwidth than comparable PNM signals.

There are a few variants of the basic PCM system already in use, but we don't really need to go into the details here. The differences are very technical and subtle. The basic operating concept remains the same.

There are several factors to be considered in any recording system. The three most important factors in audio recording and reproduction are the frequency bandwidth, the signal-to-noise ratio, and linearity.

The *frequency bandwidth*, of course, is the range of program signal frequencies that can be recorded. The *signal:noise ratio* is a measurement of the relative signal strengths of the desired program signal and undesired random noise generated by various components throughout the audio chain. Finally, *linearity* is an expression for signal accuracy. If the signal is graphed for various conditions, how much curvature will there be in the graph line? The straighter the line, the better the system's linearity.

All three of these factors are important for both analog and digital recording systems. Table 3-4 compares the various digital modulation systems on the basis of these three factors.

As you can see, PCM and PNM definitely come off the best here. In fact, PNM offers a somewhat better frequency bandwidth than PCM. But PNM is much more difficult and expensive to implement than PCM, so PCM is the commonly used modulation method. It isn't much of a compromise really. The frequency bandwidth possible with PCM is more than sufficient to handle the entire spectrum of audible frequencies. The advantages offered by PNM usually are not sufficient to justify the increase in cost and circuit complexity.

Most practical digital audio recording systems use PCM. This form of modulation is also used with CDs. The high portions of the pulses are represented as islands on the surface of the disc, while pits represent the low portions of the pulses. PCM is also being employed in large-scale communications systems, such as satellite communications and telephone systems.

Even though widespread digital recording is a relatively recent development, the PCM system is scarcely new. Pulse code modulation was invented in 1939 by A. H. Reeves. It was further developed as a modulation system for communications by C. E. Shannon in 1948.

SAMPLING FREQUENCY

Probably the most critical factor in the digital recording of analog (sound) signals is the sampling frequency. The instantaneous

Table 3-4 Comparison of digital modulation systems based on frequency bandwidth, signal noise ratio, and linearity.

Modulation Method	Frequency Bandwidth	Signal: Noise Ratio	Linearity
PAM	Good	Fair	Fair
PCM	Very good	Excellent	Excellent
PNM	Excellent	Excellent	Excellent
PPM	Good	Good	Excellent
PWM	Good	Good	Excellent

amplitude or level of the analog signal to be recorded is measured (sampled) many times each second. Each sample is then converted into an appropriate digital value. The rate at which this is done is known as the sampling frequency.

Increasing the sampling frequency increases how much data will be recorded for a recording of a given length. This causes an increase in the circuit complexity and decreases the tape economy of the recording system. Increasing the sampling frequency of a digital recorder is equivalent to using a higher tape speed on an analog tape recorder (see Chapter 2).

On the other hand, if too low a sampling frequency is used, there will not be sufficient detail in the recording to accurately reproduce the original analog signal. This is like using a slower tape speed on an analog tape recorder.

As you can see, selecting the sampling frequency is a trade-off between economy and recording fidelity, just as with selecting the tape speed for an analog tape recorder.

As you can see, selecting the sampling frequency is a trade-off between economy and recording fidelity, just as with selecting the tape speed for an analog tape recorder.

There is no theoretical upper limit to the sampling frequency that can be used in a digital recording system. A sampling frequency of 50,000,000 samples per second would work just fine, but it would definitely be overkill for recording audio signals. The recorder would be unnecessarily complex and expensive. The tape economy would be terrible in such an overdesigned recording system.

On the other hand, there is a definite lower limit to the sampling frequency used in any digital recording system. The sampling frequency must be at least twice the highest signal frequency to be recorded. If we identify the sampling frequency as $F(s)$, then no signal with a frequency higher than $F(s)/2$ can be included in the recording.

If the sampling frequency is too low, a problem known as *aliassing* is likely to occur. On playback, the circuitry can not properly decode the samples from the too-high signal frequency. These samples will be misinterpreted as noise and distortion, and possibly even a "phantom" lower frequency which was not part of the original recording signal. This effect, known as aliassing, is illustrated in Fig. 3-9.

As you can see, the designer of a digital audio recorder must

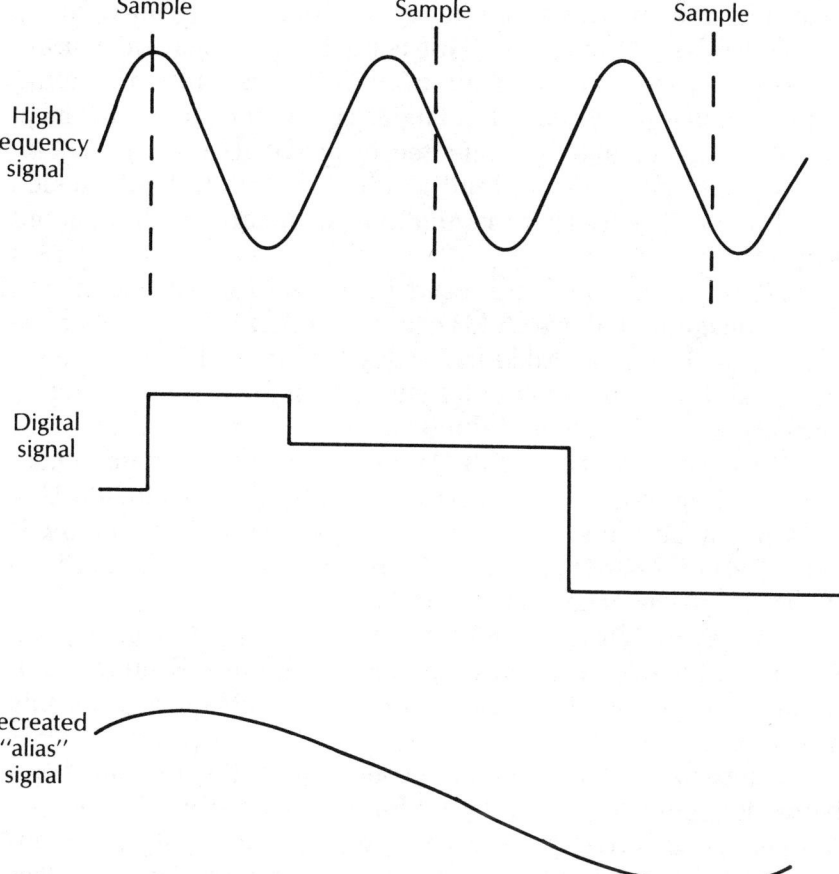

Figure 3-9 *If the sampling rate is too low, "phantom" aliassing frequencies may appear in the reproduced signal.*

carefully select the sampling frequency. In practical terms, just how high must the sampling frequency be for audio recording? Nominally, the human ear can hear frequencies from 20 Hz to 20 kHz (20,000 Hz). Therefore, the sampling frequency for a high-fidelity digital recording system should be at least 40 kHz (40,000 Hz − 20,000 + 20,000). For CDs, the standardized sampling frequency is 44.1 kHz (44,100 Hz). This gives a little extra "head room" to the system, without appreciably increasing the overall cost of the system.

If absolute high fidelity is not demanded, a recording system may get away with a somewhat lower sampling frequency, lowering the maximum recordable frequency. Many analog tape recorders are rated for frequency responses that only go up to about 15 kHz (15,000 Hz). This is not too significant a compromise. Most people cannot truly hear all the way up to the nominal human ear limit of 20 kHz. High frequency response for most people's ears tends to drop off after the age of 20 or so, especially if frequently subjected to loud sounds (including loud music). Besides, there really isn't too much musical content above about 15 kHz.

However, some sounds used in music do have significant harmonic content above 15 kHz, so something is lost by decreasing the maximum recorded frequency this way. The loss is fairly subtle, and probably won't be noticed by many listeners except, perhaps, for dedicated audiophiles.

If we set a 15-kHz (15,000-Hz) limit on the recorded signal, the sampling frequency can be lowered to 30 kHz (30,000 Hz). This might be done to improve the system's tape economy. It would not make any noticable difference in the overall complexity or cost of the necessary circuitry.

To prevent aliassing problems, we need to block off any signal content above the maximum recordable limit. Even an inaudible signal above what the ear can hear can create very audible aliassing effects.

A low-pass filter is generally used to limit the original signal before it is converted into digital form for recording. A filter is a frequency sensitive circuit. A low-pass filter blocks high-frequency components, but lets low-frequency components pass through to the output. In this application, the filter is designed to block any signal frequencies above the maximum frequency that can be reliably handled by the system's sampling frequency.

There is considerable variation in the quality of the filter circuits that can be used for this purpose. The steeper the cut-off slope, the better the results will be.

RESOLUTION

Another vitally important specification for a digital audio recording system is the resolution. The *resolution* specification is the number of bits used to represent the instantaneous amplitude of each individual sample of the audio signal to be recorded. Obviously, the higher the number of bits used, the greater the amount of detail the system can record.

If just one bit was used, there would be only two possible amplitude values: 0 and 1. This would obviously be terribly inadequate for audio recording. Remember that intermediate values are not possible in a digital system. Each bit must be either a 0 or a 1. Half bits cannot be used.

To achieve a higher resolution, there's only one possible solution. We must use more bits to define each individual sample value.

A two-bit resolution would offer four possible amplitude levels for each sample: 00, 01, 10, and 11.

Each additional bit added to the resolution doubles the number of possible instantaneous amplitude levels that can be detected for each sample. For example, three bits gives us eight possible levels: 000, 001, 010, 011, 100, 101, 110, and 111.

A four-bit resolution would offer sixteen possible recordable levels per sample: 0000, 0001, 0010, 0011, 0100, 0101, 0110, 0111, 1000, 1001, 1010, 1011, 1100, 1101, 1110, and 1111.

A five-bit system would double this to 32. Six bits would bring the total number of sample values to 64. Seven-bit resolution would give 128 possible sample values, and eight-bit resolution would give 256 possible sample values.

Practical digital recording systems use even higher resolutions than this. Usually 14 bits is considered about the minimum acceptable resolution, giving an available range of 16,384 possible instantaneous amplitude values per sample. Most modern digital recording equipment has a resolution of 14 bits, 16 bits (65,536 possible levels per sample), or 18 bits (262,144 possible levels per sample). Resolutions of 16 or 18 bits offer a very fine degree of reproducable detail in the recorded signal.

The process of breaking up the signal into discrete amplitude steps is often called quantization. The number of quantization bits is a measurement of the resolution of the digital recorder. This is a very important specification in digital recording because the number of quantization bits is directly related to the signal:noise (S:N) ratio of the recorder. The higher the number of quantization bits, the less random noise there will be in the playback signal. For example, a 16-bit system will be capable of a S:N ratio of 98 dB. Notice that this is significantly better than standard analog recording systems. For more information on signal:noise ratios, refer back to Chapter 2.

Unfortunately, the process of quantization introduces some

noise of its own into the recorded signal. Not surprisingly, this is known as quantization noise.

Quantization noise occurs because the digital recording system does not permit in-between values. Each sample must be assigned a specific and discrete whole-number value. If the actual instantaneous amplitude of the analog signal being sampled happens to fall between two adjacent quantization steps (which is likely for most samples), the conversion circuitry must round the value either up or down. Either way, some inaccuracy is introduced into the recorded signal. The recorded wave form is sometimes subtly distorted by the quantization process.

Curiously, the objectionable effects of quantization noise can be significantly reduced by intentionally adding a small amount of analog noise to the signal being digitized and recorded. White noise is used for this purpose. White noise consists of equal energy at all frequencies, and successfully masks the effects of quantization noise in most cases. This added noise signal is called dither.

ERROR CORRECTION CODES

No recording media is ever 100% perfect, of course. Some of the signal can be lost. A speck of dust, a fingerprint, or some other type of contamination can prevent the playback heads from reading a portion of the magnetic data stored on a tape.

Any practical recording tape is also inherently prone to drop-outs. A drop-out occurs when there is a spot on the tape without any magnetic particles. This happens when the binder is not strong enough (or has aged) and some of the magnetic coating chips away. A large enough drop-out is going to render that section of tape utterly useless.

In analog recording, drop-outs generally aren't too much of a problem, unless they are particularly severe. If part of a wave form cycle, or even a complete cycle or two, is not reproduced because of a tape drop-out, the human ear will tend to compensate for the gap and fill in the missing portion, as illustrated in Fig. 3-10.

In digital recording, data is much more tightly compressed on the tape. Even a very small drop-out can very easily lose two bits from a sample. In fact, that would be a fairly small drop-out.

Let's assume we have a 14-bit digital signal, and a drop-out

Error Correction Codes   93

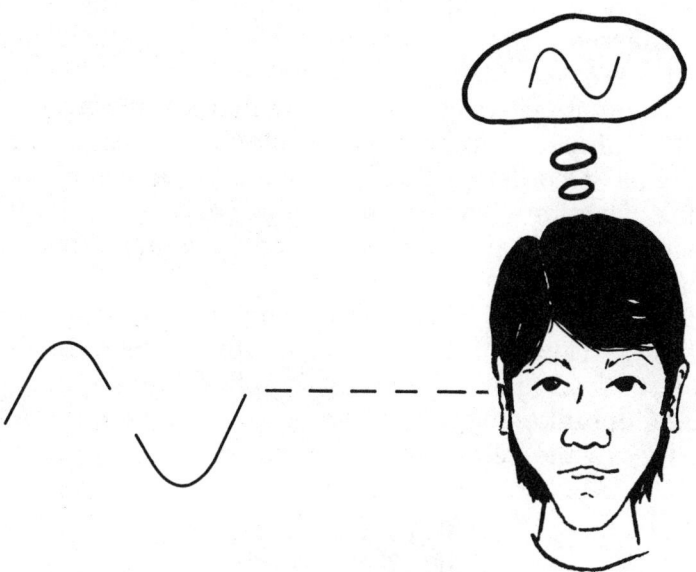

Figure 3-10 *The human ear tends to fill in gaps in analog wave forms.*

blocks the tenth and eleventh bits. The playback circuitry gets the following sample value:

00 1??0 1101 0011

There are four possibilities for the unknown bits, giving us four potential sample values:

00 1000 1101 0011
00 1010 1101 0011
00 1100 1101 0011
00 1110 1101 0011

Which is the correct value for this sample? To give you an idea of how critical this decision is, let's convert each of these potential sample values into their decimal equivalents:

00 1000 1101 0011   4307
00 1010 1101 0011   5431
00 1100 1101 0011   6455
00 1110 1101 0011   7479

As you can see, these two missing bits can make a considerable difference in the actual sampled value. If the drop-out extends across part of several adjacent samples, which is not an unlikely situation, we can be faced with a real mess.

To minimize such problems as much as possible, all practical digital recorders include some sort of error correction scheme. There are several possible approaches to digital error correction, which are often employed in conjunction with one another.

Interpolation is often used. Here, the playback circuitry looks at the values of the sample on either side of the defective sample, and assumes the missing sample value is approximately at the midpoint between the two. In our example, the playback circuitry saw the following three samples:

$$00\ 0100\ 1101\ 0011$$
$$00\ 1??0\ 1101\ 0011$$
$$00\ 1100\ 1101\ 0011$$

In this case, the sample value seems to be increasing at a rate of 00 0100 0000 per sample. The recorder's playback circuitry would probably make the reasonable assumption that the defective sample should have a value of

$$00\ 1000\ 1101\ 0011$$

or 4307 in decimal notation.

Practical interpolation systems are actually more sophisticated than this, generally using more than just the two immediately adjacent sample values to calculate the damaged sample value, but this simplified version should give you a rough idea of how the process works.

Special error correction codes are often included in the recorded data stream along with the actual sample data. In certain cases, some of the sample data is repeated in different locations along the length of the tape. The full sample value may be broken up and interwoven between several adjacent samples. The idea behind this is to spread the error effects of a drop-out over several samples at differing points, rather than concentrating them over just a few samples at the same point. This technique can

Sample 1

    A1   A2   A3

Sample 2

    B1   B2   B3

Sample 3

    C1   C2   C3

Second sample recorded as

A2 <u>B1</u> A3 <u>B2</u> C1 <u>B3</u> C2

Figure 3-11 *Breaking up a sample value and interweaving it between adjacent values can help minimize error correction problems.*

prevent a moderately large drop-out from affecting the same portion of several samples in a row. This idea is illustrated in Fig. 3-11.

A parity bit is often used for error correction purposes. This is a single bit that indicates if the current sample value is even or odd. If the parity bit does not agree with the sample value picked up by the playback head, the recorder's circuitry knows something must be wrong; although it may not be able to tell whether the error is in the sample value or the parity bit itself.

Sometimes a complex error code is recorded in addition to the actual signal data. For example, along with each sample value, the recorder may also record the difference between this sample value and the preceding sample value.

The use of error codes inevitably takes up more space on the tape, reducing recording economy, but they can vastly improve the accuracy and fidelity of the signal on playback. Like almost everything else, compromises must be made here. Somehow, we can never achieve the best of both worlds simultaneously.

Any practical analog recording media can be subject to distortion. That is, the recorded wave shape may be misshapen by the recording process. This distortion of the wave shape usually adds extra, unwanted harmonics and noise. A digital tape recorder is not bothered at all by ordinary distortion in the recorded signal, unless it is extremely severe, resulting in lost data similar to drop-outs.

In a digital recording, the recorded signal is always a pulse

96   Basics of Digital Recording

wave, which has jut two levels—high and low. As long as the playback circuitry can distinguish between these two levels, the signal can be easily and accurately reshaped. If the signal at instant x is above z volts, the wave-shaping circuitry puts out a high pulse. Then, if the signal at instant y is below z volts, the playback circuitry treats it as a low pulse. It is very easy for the wave-shaping circuitry in the playback recorder to accurately recreate the original pulse wave signal, unless there are severe distortion problems in the recording.

PERFECT COPIES

Any analog recording system degrades the signal fidelity each time it is rerecorded. If you make a copy of a recording, the distortion in the playback signal will necessarily be increased. If you make a copy of a copy of a recording, the distortion problems will always be worse with the later generation copy.

At best, you can expect the distortion to increase by at least about 0.5% on each new generation of recording. This does not include any distortion effects introduced by other elements of the playback system (the amplifier and, especially, the speakers). Audiophiles can usually hear distortion levels of 1%. A distortion level of 3% will be clearly audible to almost any listener.

This recording distortion occurs because the actual recording media (vinyl discs or magnetic tape) always includes some built-in and unavoidable distortion elements. For example, in a magnetic tape recording, some of the magnetic particles (especially if they are relatively large) may not completely line up to the magnetic field of the recording head properly. A certain number of the magnetic particles will remain randomly arranged, particularly at low signal amplitudes. This will result in a poor signal:noise (S:N) ratio on playback.

Analog discs and magnetic tape are also inherently nonlinear. They do not record everything equally. They are not capable of recording and reproducing a signal with total accuracy.

As an example, let's consider the issue of frequency response. If the recording tape tends to boost frequency X slightly, but attenuates frequency Y somewhat, this effect will be exaggerated. For purposes of discussion, let's say the intended amplitude of both these frequency components is 1 unit. (It doesn't matter what that unit of amplitude measurement is. We're dealing with relative comparisons rather than absolute measurements.) We will assume that the frequency distortion for both frequency X and frequency Y is 5%. (This is actually a bit extreme, but it will do for purposes of discussion.)

When the first recording is played back, the actual amplitude of frequency X is 1.05 units, while that of frequency Y is 0.95 unit. Now, if we make a copy of this recording, frequency X will be boosted another 5% and frequency Y will be attenuated an extra 5%. When this copy is played back, the actual amplitude of frequency X is 1.1025 units, while that of frequency Y is 0.9025 unit. On a third generation recording, frequency X will be up to 1.157625 units and frequency Y will be down to 0.857375 unit.

As you can see, the cumulative distortion effects can very quickly build up to totally unacceptable levels when multiple generations of tape are duplicated. Such problems are inherent and inescapable with any analog recording system. Advanced technology can sometimes minimize the adverse effects somewhat, but they can never be eliminated. The higher the generation number (number of copies) of a recording, the lower the actual fidelity of the final tape is going to be.

This multigeneration tape duplication problem does not occur in a digital recording system. A certain amount of quantization distortion may be introduced on the original master recording, but the effect is not cumulative. It does not get worse with additional copies. This is because digital signals are inherently much less ambiguous than analog signals. As long as the signal is not converted back into analog form, no further distortion will be added, no matter how many copies are made.

Once the signal has been properly digitally recorded, the data pulses are reshaped on playback. These reshaped pulse waves are then rerecorded onto the copy, but all of the encoded data remains exactly the same (excluding any uncorrectable data). There is no increase in the signal distortion when a copy of a copy is made digitally.

Theoretically, hundreds or even thousands of copies may be made of a digital recording without significant degradation of the signal fidelity and reproduced sound quality. As long as the digital data (the actual encoded 1s and 0s) are not corrupted, you can't tell the first-generation master recording from a fifth-generation copy. Obviously, this is the great advantage of digital audio recording, especially when multiple mixing stages are involved.

MULTIPLEXED SIGNALS

One of the most powerful advantages of PCM is the ability to multiplex signals. By multiplexing, more than one signal can be imposed onto a single PCM carrier. This process is also sometimes known as time sharing. The process of reseparating the multiplexed information is called demultiplexing.

The two channels (right and left) in a stereo recording can be multiplexed, so that there is absolutely no crosstalk between them. That is, none of the right channel signal leaks over into the left channel, or vice versa.

A crude form of multiplexing is used on stereo LPs, as described in Chapter 2. The right and left channel information is cut into a single groove at opposing angles. Both signals are picked up by a single stylus and are separated (demultiplexed) for true stereo sound reproduction. The crudeness of this simple form of multiplexing is the reason stereo separation on records is only fair. Tape and digital media typically offer much better stereo separation than the old vinyl disc.

Other multiplexed signals in a typical digital audio recorder include the error correction codes (discussed earlier in this chapter) and various identification codes. "Hidden" code can be used to tell the recorder what tape is currently being played, and the present position on the tape. Position indicators can be useful for rapidly locating a specific selection recorded on the tape.

REQUIREMENTS FOR A DIGITAL AUDIO TAPE RECORDER

A practical digital audio tape recorder must be designed to meet certain minimum requirements. Many of the specifications in a digital design must be far more stringent than in an analog system. Other specifications become totally irrelevant as we go from analog to digital recording.

A digital tape recorder demands a very large bandwidth, often as high as 30 times the bandwidth of a high-quality analog tape recorder. This extremely wide bandwidth is necessary because so much data must be so tightly compressed to give reasonable tape economy. The data recorded by a PCM recorder is extremely dense.

Generally speaking, higher grade tape is required for digital recording. The magnetic particles should be as physically small as possible to accomodate the extremely high recorded data density. A good-quality binder is also required to reduce the tendency for some of the magnetic particles to chip off in spots, creating drop-outs. Drop-outs are generally much more of a problem in digital tape recording than in analog tape recording. Error correction codes can compensate for relatively small losses of data, but no error correction system can be expected to perform miracles. If too much data are lost, the recorded signal cannot be successfully reproduced.

In a digital tape recorder, only pulse waves (representing 1s and 0s) are actually recorded onto the tape. The pulse wave form can easily be reshaped during playback; thus high linearity of the recording media (the tape) is not as essential as it is in an analog system. In an analog tape recorder, poor linearity will result in possibly severe distortion of the recorded signal. This is not a problem with a digital tape recorder.

All practical analog tape recorders use some kind of high-frequency ac bias to improve the recording linearity. The concept of the record bias signal was described in Chapter 2. This linearity improving record bias signal is not required in a digital system. PCM recorders do not include a bias oscillator in their circuitry.

In a PCM recorder, the data bits for one channel can be dispersed onto a number of tracks to preserve data integrity. Data lost because of a drop-out affecting track A may be recovered from the data for that channel stored on track B. Data tracks are often broken up and interwoven on the tape, as explained earlier in this chapter. This minimizes the probability of catastrophic data loss.

Multiplexing is also possible with a PCM tape recorder. Multiple channels can be recorded onto a single digital track without interfering with each other in anyway. On playback, the multiplexed channels can be fully separated (demultiplexed) without any measurable crosstalk between them.

There are two basic mechanical designs used for digital tape recorders: stationary-head recorders and rotating-head recorders. Standard analog audio tape recorders are all of the stationary-head type. The physical position of the record or playback head is fixed. The tape is pulled across the stationary head at a specific rate. A typical stationary-head arrangement is illustrated in Fig. 3-12.

In a rotating-head recorder, two (or more) heads are placed

100 Basics of Digital Recording

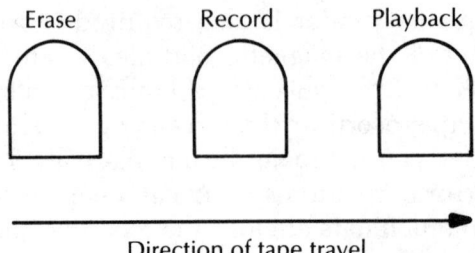

Figure 3-12 *This is the arrangement used in a typical stationary-head digital tape recorder.*

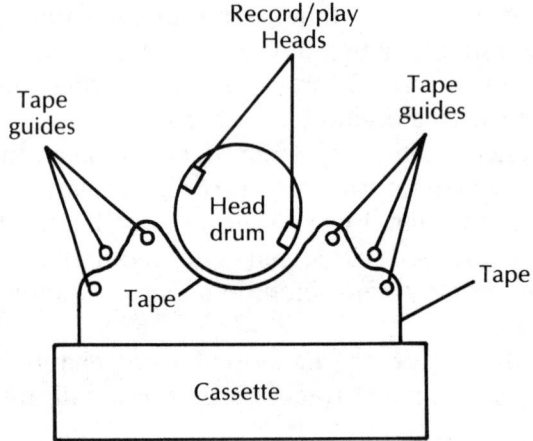

Figure 3-13 *This is the arrangement used in a typical rotating-head digital tape recorder.*

on a rotating drum. The tape is partially wrapped around the head, as shown in Fig. 3-13. The tape is slowly pulled past the head drum as the head drum rapidly revolves. This greatly increases the effective tape-to-head speed, allowing higher data density to be recorded on a given amount of tape. The effective tape-to-head speed is a combination of both the actual speed of the tape's movement past the head drum and the speed at which the heads are rotated past the tape in the opposite direction. The absolute speed of the tape is not as important as the speed of the tape relative to the head. The recorded signals don't care whether the tape is moving and the head is stationary, whether the tape is stationary and the head is moving, or if both are moving. By moving both the tape and the head (by rotation), the greatest relative tape-to-head speed can be achieved with a minimum effort. High data density and good tape economy can both be achieved with this system.

Rotating heads are used in VCRs because of the large amounts of information that must be recorded to store and reproduce video signals.

STATIONARY-HEAD RECORDERS

At first glance, it might seem that a stationary-head machine would logically be far simpler and less expensive than a rotating head recorder. In reality, for digital recording this is not the case. It is actually quite difficult to design and build a functional stationary head digital tape recorder.

The problem is moving the tape past the head at a sufficient speed to permit adequate data density in the recording. Digital data requires very high storage density on the tape. In other words, it takes up a lot of room. The tape has to be moved past the recording head very rapidly to provide enough room for all the necessary digital data to fit onto the tape.

A rotating-head machine can easily achieve effective tape-to-head speeds in the neighborhood of 10 m/s (2,540 IPS) or so, without consuming enormous quantities of tape. (The way this works will be discussed in the next section of this chapter.)

It just isn't practical to whip tape past a stationary head anywhere near that fast. Even if you could physically pull it off, such an attempt would obviously result in extremely poor tape economy. A single 2-minute recording would require 1,200 meters (25,400 feet) of tape. That is unquestionably a lot of tape. To provide a little perspective, consider that a standard C-90 audio cassette, which has a playing time of 45 min/side, contains just a bit over 420 feet of tape—a very considerable difference in tape economy. If a recording of a complete symphony was made at such speeds, the resulting tape would probably be difficult to even lift.

Besides the obviously unacceptable tape economy, such high speeds would place extreme physical strain on the tape itself, the recorder's motors, and the tape guides. Tape stretching and breakage would almost certainly be a constant problem with such an impractical system.

The maximum reasonable tape speed for a stationary head tape recorder is considered to be about 76 cm/s (193 IPS). This still isn't great in terms of tape economy. A 2-minute song would take up 1,930 feet of tape, and 30 minutes of recorded music (probably the minimum practical length for a commercial recording medium) would require no less than 28,950 feet. Tapes for such a machine would be very bulky and expensive, and the tremendous tape speed would make for only a fair-quality recording.

A slower tape speed (with respect to the head) means that there is less room to store data on the tape. Data must be much more tightly compressed, leaving far less margin for error. The tolerances for the circuitry in such a recorder must be extremely tight and precise. The tape could barely hold the minimal signal data. There would be little or no room on the tape for any control data or error correction codes. Most of the special advantages and potentials of digital recording would be lost in this case.

The basic arrangement of the heads in a typical stationary-head digital tape recorder is shown in Fig. 3-14. Notice that this arrangement is quite similar to the arrangement of the heads on a standard analog tape recorder, except there are two playback heads, one record head, and no erase head. No, this is not an error. That is the way this type of digital recorder is set up.

The main playback head is mounted after the record head, allowing the signal to be monitored immediately after it has been recorded. The second playback head is mounted just before the record head on the tape's path. This second playback head is

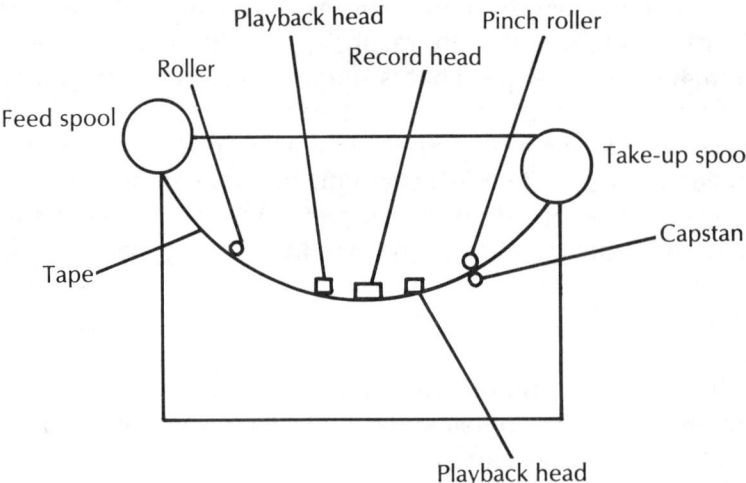

Figure 3-14 *A stationary-head digital tape recorder is mechanically similar to an analog tape recorder.*

used to monitor previously recorded material for matching synchronization signals while dubbing new material onto the tape on a different channel. In an ordinary analog tape recorder with sync capabilities, part of the record head is used as a temporary playback head for such synchronization purposes. For various technical reasons, this just won't work with the digital tape recorder, and a second, specialized playback head is required to do the trick.

Internal signal delay circuitry ensures that the reproduced signal from this secondary playback head is in proper synchronization with the new signal being laid down by the record head. To correctly reproduce the recorded digital signal, the recorder's circuitry must be synchronized with the tape signal. A special sync signal is recorded along with the code for the audio signal for the audio signal for timing purposes. The ability to erase part of the previously recorded signal could obviously play havoc with the system's internal synchronization, so the erase head is omitted from the stationary-head digital tape recorder. Digital tapes must be bulk erased. Errors are normally corrected via electronic editing tricks.

The stationary-head tape recorder is unlikely to ever become a consumer item. This type of machine is necessarily large, complex, delicate, and extremely expensive, and it has lousy tape economy even under the very best of circumstances. It may be useful in some professional studio work, but generally speaking, the rotating-head digital tape recorder seems much more promising as a potential commercial product.

ROTARY-HEAD RECORDERS

In a rotary-head tape recorder, two (or sometimes more) record and playback heads are mounted on the rim of a rotating drum. The tape is threaded through the machine in such a way that it is partially wound around the circumference of the head drum, as illustrated in Fig. 3-15.

In most practical rotary-head systems, the tape is contained in some sort of enclosed cassette. This protects the tape from fingerprints and other possible contamination. The cassette housing also makes using the tape much more convenient. The user simply inserts the tape cassette into the appropriate slot in the recorder. An automated mechanism of some sort pulls an

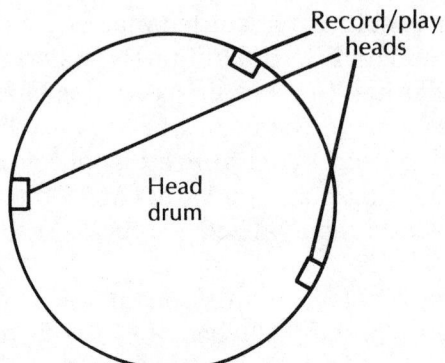

Figure 3-15 *Multiple record/play heads are mounted around the surface of the rotating head drum.*

appropriate amount of tape out of the cassette and wraps it around the drum head. This is not merely a nifty labor-saving device. The threading requirements in rotary head machines are usually fairly complex and demand absolute precision. An automated loading system ensures that the tape is threaded properly, minimizing the chances of tape breakage, jamming, or bending of delicate tape guides. A mistake, which could be very possible with manual loading, would have disastrous results.

Another reason for the automated loading mechanism is that the tape's tension is very critical. It would be difficult, or at least awkward, to set the proper tension manually, but a machine can do the job very easily and neatly.

Finally, by using automatic loading of the tape, the user never has any reason to touch the tape itself, reducing contamination risks.

In operation, the tape is pulled across the head drum at a fairly low speed. At the same time, the head drum is rapidly rotating in the opposite direction. This gives a very high effective tape-to-head speed. A lot of data can be crammed onto a relatively small amount of tape.

Usually, the heads are mounted in the drum at an angle. This creates a series of diagonal tracks across the tape, as illustrated in Fig. 3-16, allowing for maximum data density to be recorded onto the tape.

Often the heads mounted on the opposite sides of the rotating drum are angled in opposite directions. The head angle is called the *azimuth*. Contrasting azimuths minimizes any potential overlap or crosstalk problems from adjacent tracks.

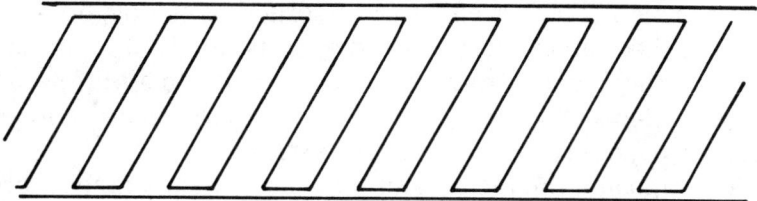

Figure 3-16 *The data is recorded as a series of diagonal tracks across the width of the tape.*

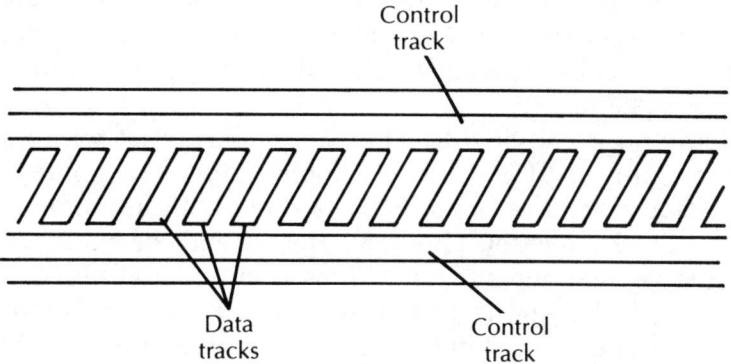

Figure 3-17 *In some systems, additional control tracks are run along the edges of the tape.*

Because of the narrow, diagonal tracks along the length of the tape, a digital tape from a rotary-head recorder cannot be physically edited using cut and splice methods like with ordinary analog tape recordings. All editing with digital tape recorders is done electronically. The data from one (or, in some cases, more than one) recorder is rerecorded onto a second recorder.

If a separate erase head is used, it is usually of the stationary type, just as in an ordinary analog tape recorder. There is no particular need for high tape-to-head speed for erasure.

A separate control track is normally recorded along with the music data to keep the playback circuitry properly synchronized. Often a separate set of (stationary) record and play heads are used for this purpose. The control track is recorded along a continuous narrow strip along the top or bottom of the tape, as shown in Fig. 3-17.

PCM CONVERTERS AND VCRS

A VCR also uses a rotary head system, because video signals also require a very wide bandwidth and high recording density. The video signals in modern VCRs are recorded in analog, not digital form, but the high frequencies and wide bandwidth required for an analog signal make the recorder a necessity. In addition to the rotating video head drum, separate stationary record and play heads are used for the (analog) audio track.

The tape and the record and play heads in any tape recorder don't really care what signals are placed onto or are read from the tape. The recorder heads just see voltages and magnetic fields. The meaning of these signals and how they are used is determined by electronic circuitry. As long as the heads and the tape can handle the signal frequencies used, there is no effective difference between analog and digital signals at this point in the recording chain.

Early on, some bright technician realized that with a little extra circuitry a VCR could be "fooled" into functioning as a digital audio tape recorder. This is done with an add-on unit called a PCM converter. This converter accepts the analog sound signal to be recorded and transforms it into digital (PCM) form. Then the PCM signal is "disguised" to look like a video signal, using 30 frames per second, along with standard video sync pulses. This signal is fed into the VCR's inputs to be recorded. During playback, the PCM converter works in just the opposite way.

It made perfect sense to use a converter with an existing VCR. There was no good reason to reinvent the wheel for this application; the VCR would do the job just fine. Since the PCM converter didn't need to include the actual record and playback circuitry or mechanisms, the price tag could be reduced. Even so, PCM converters were quite expensive.

Several home PCM converters were marketed in the 1980s, including the Sony F-1 and the DBX 700, among other models. This format never really caught on with the general public. The PCM converters were very expensive and rather complicated to use. Also, the average consumer is not used to thinking of his VCR as an audio tape deck. The appeal was solely to the most dedicated (and wealthy) of audiophiles, and producers of professional recordings.

Still, the idea of digital audio recording offered a lot of appeal. Manufacturers felt that in the right, user-friendly format, digital recording could be a viable product. It could conceivably spark a full-fledged revolution in the audio industry. It was clearly an idea whose time had come.

The trick was to find a suitable easy-to-use and relatively inexpensive format for digital recording. First came a digital playback-only medium—the CD (discussed in detail in the following chapter). The CD has done very well in the marketplace. Many experts were surprised by the degree of its success.

Now, DAT appears to be a very practical approach to consumer-level digital recording. Details of this format will be discussed in Chapter 5. Of course, with the development of DAT, separate PCM converter and VCR combinations are rendered unnecessary and impractical. A DAT recorder can do the whole job with a single, dedicated machine. A DAT recorder costs less than a PCM converter/VCR combination and is a lot less fuss and bother to use.

It is interesting to note that the actual record and play mechanism of a DAT recorder is very similar in design to a standard analog VCR.

THE ADVANTAGES AND DISADVANTAGES OF DIGITAL RECORDING

Digital recording of audio signals offers many significant sonic advantages over traditional analog recording techniques. The improvements of digital audio recording include

- No surface noise or tape hiss,
- Very low distortion,
- Very wide and realistic dynamic range, and
- Very flat frequency response.

Of course, no technology is perfect. Digital recording of audio signals has certain limitations and disadvantages of its own. The most significant of these disadvantages are

- High equipment cost,
- Quantization noise,
- Possibility of aliassing effects, and
- Impossibility of manual edits.

The high cost is likely to come down as factories get into gear to supply the new technology. Quantization noise and aliassing problems can usually be reduced to negligible levels by a good circuit designer.

The editing problem, however, is almost certainly here to stay. Because of the way data is packed onto the tape in a digital recorder, manual cut-and-splice edits like those used with analog tapes, just won't work. On the other hand, electronic edits (performed by dubbing from one or more tape decks onto another recorder) are much more practical and efficient with digital recordings. There is no degradation of the signal quality, and numerous special effects can be added to the signal, if desired.

On the whole, the advantages of digital audio recordings seem to outweigh the disadvantages. This will be even more true when digital recording equipment prices come down.

❖ 4
Compact Discs

THE FIRST DIGITAL RECORDING MEDIUM TO REALLY CATCH ON with the general public is the compact disc (CD). This is a playback-only medium. It is roughly the digital equivalent of the old vinyl LP.

The CD is a nonphotographic optical storage medium that is scanned by a laser to pick up digitally encoded data. This definition certainly sounds like an awful mouthful, but it isn't too complicated if we take it step by step.

Nonphotographic means that no photographic negatives or positives are used to make the disc. The CD is not related to cameras or film. No pictures are involved (unless the digitally encoded data represents a video image). This distinction is made because a photographic connection might be mistakenly assumed from the "optical" part of the description.

Optical means having to do with sight or light. A CD is read by light waves.

Storage medium means nothing more than this is a system for recording and storing information.

Scanned by a laser refers to the way the encoded data is detected when the disc is played. A *laser* is just a very coherent and tightly focused light source. The use of the laser is the reason the CD is known as an optical medium.

Finally, the signal detected by the laser is in the form of *digitally encoded data*; that is, the information on the CD is in the form of binary 1s and 0s, or digital pulses.

The digital data on a CD can represent anything at all. There is nothing inherent in the design of the CD to limit it to audio recordings. For example, there have been experiments with storing video images on CDs. CDs are also used to permanently store computer data. In this book, however, we will concentrate solely on audio recording.

In many ways, the CD is the digital equivalent of the old vinyl LP, but there are some important differences between these two recording media. Of course, the most important difference is that a standard LP record is an analog recording, and a CD is a digital recording.

There are other differences too. A standard LP record is played by a stylus physically riding in the grooves etched into the surface of the disc. This physical contact contributes to the wear and tear on the record. It is worn down a little each and every time it is played.

A CD, on the other hand, is played by reflecting a beam of light from a small laser. The data on the disc is read optically rather than mechanically. Because there is no actual physical contact with the surface of the disc, there is no wear and tear on the CD during normal playing. The potential lifespan of a CD is theoretically infinite.

The information on a standard record is in the form of a squiggly set of grooves etched into the disc's surface. Data on a CD is in the form of a series of pits (indentations) and islands (raised areas). A protective transparent plastic coating is layered over the encoded data to protect the surface of the disc from possible scratches and other damages.

A vinyl record is easily scratched, and every little scratch adversely affects the reproduced sound. A CD is relatively immune to scratches. The recording surface itself cannot be scratched because it is protected by the outer plastic coating. Small scratches in the plastic coating are effectively ignored by the CD player. Severe surface scratches, however, may affect the pattern of light beams reflected back from the disc's surface, which could cause a misreading or loss of data. This is relatively rare, but it can happen.

Both an LP and a CD are spun on a turntable while being played. A standard record spins fairly slowly—just a couple dozen complete revolutions per minute. The speed of revolutions is constant throughout the entire disc. The recorded information is read from a spiral starting at the outer rim of the disc and circling inward.

A CD spins much, much faster than an LP during playback. The revolution rate is several hundred complete turns per minute. The data starts near the center of the disc and spirals outward toward the disc's outer rim. The disc spins slower when the laser is focused near the center of the disc, and faster when the laser is focused near the outer edge.

A standard LP record is recorded on both sides, but a CD is recorded only on one side; but a CD can hold more music than an LP. A CD can hold more music on its single side than the average LP can hold on both sides.

IMPROVED RECORDING

The high-fidelity audio scene has always been one of steady evolution and improvement. By the late 1960s, a number of Japanese audio manufacturers decided it was time for a major step forward. These pioneering companies included Columbia (Denon), NHK, Nippon, and Sony, among others. They developed two basic approaches to improving the sound from the vinyl LP. They recommended direct-to-disc recording and digital recording.

Direct-to-disc recording was something of a throwback to the old 78-rpm and Edison cylinder days. No master tape or editing was used. As the music was originally played, it was cut directly into the grooves of a disc. A limited number of copies could then be molded from this "mother" disc. Once the mother disc was worn out, no more copies could be made.

The musicians had to perform everything on an album side straight through, with no mistakes or pauses. There was no possible way to do any editing with direct-to-disc recording.

While direct-to-disc recording echoed the early, primitive beginnings of the recording industry, it was not really an exercise in pointlessness. The reproduced sound of an LP could be improved by direct-to-disc recording. Normally, LPs are made from tape recordings. Any tape recorder adds some noise and distortion to the signal it is recording. By eliminating this master recording stage of the reproduction chain, a direct-to-disc LP could offer some improvement in audio quality.

Unfortunately, there wasn't enough of an improvement to entice the average audio consumer. Direct-to-disc albums appealed to audiophiles, but their higher than normal price tags discouraged other customers. The higher prices for direct-to-disc recordings were necessary because ordinary mass-production techniques couldn't be used.

Perhaps it is just as well that direct-to-disc LPs never did really catch on with the general public. A record producer could run into real problems with a best selling hit record in the direct-to-disc format, since there was a definite limit to the number of copies that could be struck from the original mother disc. There was no way to meet a very high demand for a particular direct-to-disc recording if it ever came up.

The other idea suggested by those pioneering Japanese audio companies seemed more far fetched at the time, but it turned out to be far more promising. This was the concept of digital recording. By converting analog music into a string of numbers (digital data), many long-standing noise and distortion problems of analog recording techniques could be bypassed.

The idea of digital recording had been suggested from time to time in the past, but the existing technology wasn't sufficient to actually put the idea into practice. It was the development of video tape recorders that made digital recording a viable possibility.

Digital data demands an extremely large recording bandwidth, as does the information required to record a moving video image. High relative tape-to-head speeds are necessary for both video recording and digital recording. Once practical video tape recorders were devised, it didn't take too long to adapt the same basic principles to accommodate digital recording of audio signals.

At first, digital recording was only used to create the master tape. This was inevitable, since early digital recording equipment was large, bulky, tricky to use, and very, very expensive. It was nowhere near ready to be a consumer-level product.

However, a digital master tape could eliminate many of the noise and distortion problems that limited standard LPs almost as well as direct-to-disc recording could. The original analog music signal was converted into digital form and recorded. The recording could then be edited or mixed with other recordings until a complete, high-quality digital master tape was ready. The digital signals from the master tape were then converted back into analog form to be cut into the grooves of an LP record, with a noticeable improvement in the sound quality.

Most experienced listeners can tell the difference between a direct-to-disc recording (or a digitally mastered LP) and a standard analog LP, but few people can distinguish between a direct-to-disc LP and a digitally mastered LP. Digital mastering offers the added advantage of permitting stop and start recording, mixing, and recording. Also, there is no limit of perfectly identical copies that can be made from the original digital master tape.

The producers of the earliest digitally mastered records in the early 1970s didn't even bother to advertise the fact. But they soon learned that digital was a hot "buzz word" for sales.

Today digital mastering is used for the majority of recordings of classical music. It is somewhat less commonly used for recordings of popular music, where the advantages are of less significance. However, digital recording is now being used more and more frequently even in the area of popular music records.

These digitally mastered records are sometimes called "digital LPs," but that is a bit misleading. The LP itself remained strictly an analog recording medium. Only the recording's master tape is in digital form.

There were some experiments conducted in Germany in the 1970s to create a truly digital phonograph. A special cutting lathe was required to make such discs, and the required data density demanded that the discs be rotated at very high speeds, severely limiting the available recording time.

Of course, the digitally cut record was completely incompatible with any existing turntables and record players. A specialized (and expensive) turntable and stylus assembly was required to play these discs.

Not surprisingly, these experimental digital phonographs were soon abandoned. They were never marketed as a consumer product. A nice idea, perhaps, but hardly a successful attempt. Still, there was definitely strong commercial appeal in the possibility of a true digital consumer recording medium.

LASERS

A CD does not use a stylus, like a traditional analog record player. Instead, the information stored on the disc is retrieved by a beam of laser light.

The idea of reading information off a disc with a laser light beam is inherently appealing. Normal mechanical styli are inevitably destructive. Because of their physical contact with the record's groove walls, a little damage is done by the stylus each and every time the record is played. If the record could be played by shining a beam of light on it, instead of using a stylus, record life could be substantially prolonged, because there is nothing physically eroding the information etched into the groove walls.

A laser of some sort is pretty much essential for this type of job. Ordinary light cannot be focused finely enough to detect the microscopic variations in the record's surface. A laser is simply a highly coherent, tightly focused light source. It might be worthwhile to take just a moment here to consider just what these terms mean.

Ordinary light is largely incoherent. This means, among other things, that a number of differing frequencies are simultaneously generated by the light source. Light comes in different frequencies, just like sound. The frequency of the light waves determines the color of the light; ranging from red at the lowest visible frequencies, to violet at the highest visible frequencies. Light frequencies too low to be seen by the human eye are called *infrared*, and at the opposite extreme, light frequencies too high to be seen by the human eye are known as *ultraviolet*. *White light* is actually a combination of all light frequencies. It is roughly similar to *white noise*, which is a randomized combination of all audible sound frequencies.

A laser, being coherent, is monochromatic. This mouth-filling term just means "all one color." Only a single frequency of light is contained within the laser beam. To make a sonic comparison, it is rather like a *sine wave*, which is a pure, single frequency wave form with no harmonic content at all.

The term *coherence* means more than monochromatic. Besides just a single frequency, all of the waves in a laser are in phase. In ordinary light, waves are usually out of phase with one another.

Phase refers to the beginning and end point of each cycle. For simplicity here, we will just consider two simultaneous waves. Since both waves have the exact same frequency, their cycles last exactly the same amount of time. If both waves begin their cycles at the same instant, as shown in Fig. 4-1, they are in phase with one another. Since the cycle length is equal for the two waves, both will reach the end of the cycle at the same time, and will begin the next cycle in synchronization with each other. In-phase waves reinforce each other. When one goes up, the

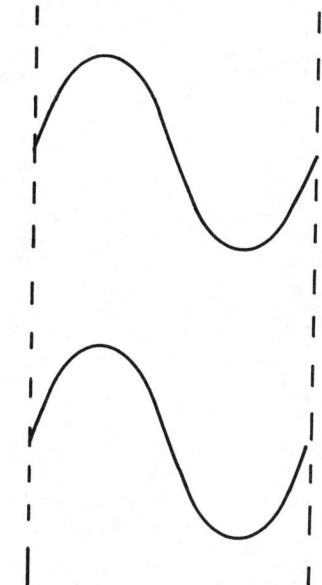

Figure 4-1 When two waves begin and end their cycles at the same time, they are in phase.

other goes up, and when one goes down, the other goes down. The waves are always "pushing" or "pulling" in the same direction.

If two waves do not begin their cycles at the exact same instant, they are out of phase, as illustrated in Fig. 4-2. When one wave is "pushing", the other wave may also be "pushing," or it may be "pulling." If both waves happen to be going in the same direction, they will reinforce each other at that point. However, any time during their cycles that the two waves are moving in opposite directions, they will partially or completely cancel each other out.

The difference between the starting points of the cycles for the two waves is called the *phase shift*. If two waves are in phase, they have a 0-degree phase shift. There are 360 degrees to each complete cycle, so two repeating waves that are 360 degrees out of phase are effectively in phase. Limiting ourselves to whole numbers (just for convenience in discussion), out-of-phase waves can exhibit wave shifts from 1 degree up to 359 degrees.

A special case occurs at the midpoint of the cycle. When two waves are 180 degrees out of phase with each other as shown in Fig. 4-3, they are always moving in opposite directions. When

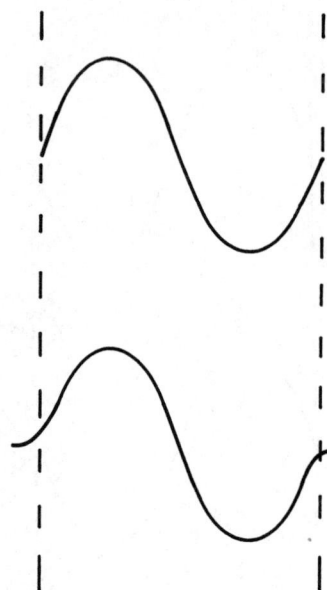

Figure 4-2 When two waves begin and end their cycles at different times, they are out of phase.

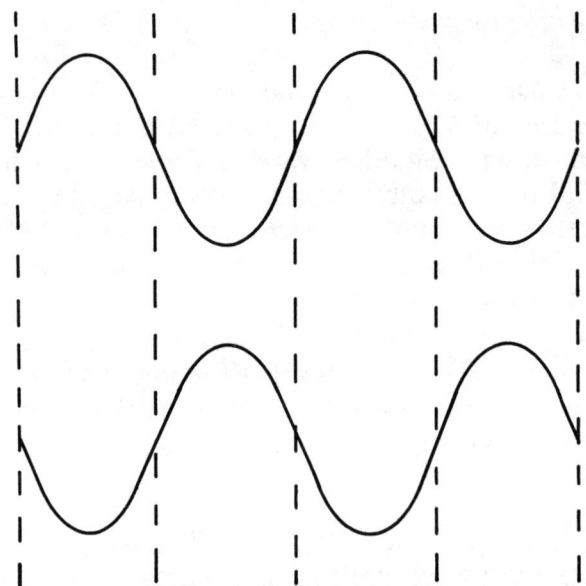

Figure 4-3 When two waves are 180-deg out of phase, they cancel each other out.

one is going up, the other is going down by an equal (opposite) amount. The two waves fight against each other and are equally matched, so they cancel each other out completely.

Of course, when waves of differing frequencies are combined (as in the light from an ordinary light source), the waves will go in and out of phase on different cycles. So, when we say laser light is coherent, we mean that all of its component waves are at the same frequency and are all in phase with one another. This is one of the main reasons a laser beam can pack so much power.

The other reason for the immense potential power of a laser is that the light is emitted in a very tight beam. Ordinary light is prone to scattering. It goes in every direction, unless something blocks it.

If ordinary light is sent out in a beam, such as from a flashlight, as shown in Fig. 4-4, the beam will rapidly spread out over

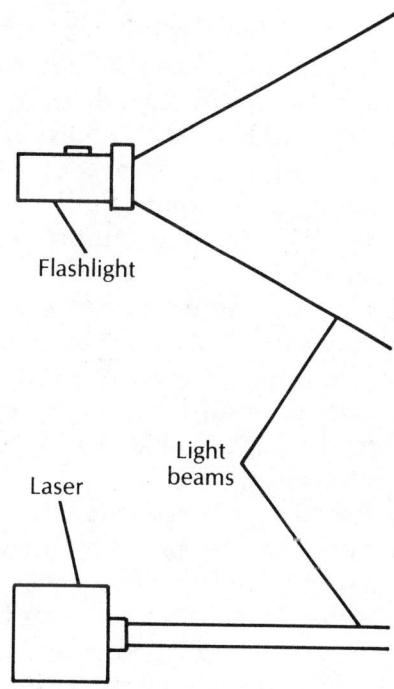

Figure 4-4 *A beam of ordinary light rapidly spreads out over a wide area.*

a wider and wider area. Since the energy is spread out, it gets progressively weaker.

A laser beam, on the other hand, exhibits very little spreading. If a laser beam the width of a standard pencil was sent all the way to the moon, by the time it got there, it would only be spread out to about the size of a dime.

In a CD player, we aren't really interested in the concentrated power of a laser. In fact, the laser used in any CD player is a very, very low-power unit. A laser is used in this application because of its purity and precision. The coherence and beam-like nature of laser light makes it very predictable and controllable, which is precisely what is needed in this application.

In a CD player, the laser beam shines onto the surface of the CD. The reflected laser light from the disc is then retrieved and measured to reproduce the recorded data. We will go into more detail on just how this is done later in this chapter.

VIDEODISCS AND CDS

We have already seen that digital audio tape recording has much in common with analog videotapes. Similarly, CDs utilize the same basic technology as optical videodiscs.

In 1972, Phillips and MCA more or less simultaneously introduced remarkably similar prototypes for a laser-scanned videodisc. These prototypes were quite similar to the Pioneer Laserdisc system that is still in use and gaining popularity.

The Pioneer Laserdisc is 12 inches in diameter (the same size as a standard vinyl LP). In the player, the disc is rotated an average speed of 1800 rpm. The actual rotation speed varies depending on where the laser is positioned on the disc. Unlike standard audio LPs, the Laserdisc is played from the inside out. The recorded information starts near the center of the disc and ends up near the outside rim.

The disc's surface is highly reflective. Encoded data, in the form of small holes or "pits" separated by small "islands," is detected by the differences in the way they reflect the laser beam.

The video Laserdisc is not a digital recording medium. The signal is in analog form. It is encoded by FM (frequency modulation). The length of the individual pits, as is the case with all analog media, can be any of a continuous and infinite range of intermediate values.

Once such a laser-scanned optical disc could be made functional for the wide bandwidth required for analog video signals, it didn't take much technological tinkering to modify the system for the wide bandwidth required for digitally encoded audio.

While the video Laserdisc and the audio CD are very similar on the theoretical level, they are totally different and incompatible on the practical level. Completely different electronic circuitry is required to play the two types of optical discs. This should not be particularly surprising, since the video Laserdisc is an analog recording medium and the audio CD is a digital recording medium.

Recently a few manufacturers have introduced some "all-in-one" laser disc players, which can play both video Laserdiscs and audio CDs. Inside such a multifunction player, you'll find two complete sets of decoding circuits. In effect, there is a video Laserdisc player and a separate audio CD player sharing a common case, turntable, and laser assembly.

THE COMPACT DISC

The physical dimensions of a CD must meet certain exact standards. A nonstandard disc cannot be played on standard CD players. The precise mechanisms in a CD player demand precise adherence to the standardized dimensions.

The CD measures approximately 4.75 inches (exactly 120 mm) in diameter. The hole drilled in the center of the CD has a diameter of 15 mm. The thickness of vinyl LPs varies a great deal, depending on the specific materials and molding and pressing techniques used by the manufacturers. A CD, on the other hand, is always 1.2 mm thick.

A standard CD can hold up to a maximum of 74 minutes of music, although some special recording tricks can sometimes be used to achieve a somewhat longer playing time. Most CDs on the market have playing times considerably shorter than the standard's maximum. Most CD releases have playing times of less than 1 hour. A shocking number of commercial CD releases are less than 30 minutes, effectively wasting more than half of the CD's capability. The average length for CD releases today seems to be in the area of 45 to 50 minutes.

There were attempts to market a smaller CD format using 3-inch discs. Some CD players could play both disc sizes directly, but most machines required that a plastic adapter be fitted onto the 3-inch CDs. Since this format was supposed to be the CD's equivalent of the vinyl single, the parallel to the adapters that usually had to be used with 45-rpm singles seems appropriate. Interestingly, the adapter worked in just the opposite way. For the 45-rpm vinyl disc, the adapter was fitted into the larger than normal center hole to fit the standard spindle. The center hole on a 3-inch CD is the same as that in a standard, full-sized CD. The adapter fits around the disc's outer rim to increase its diameter. This curious difference in adapters can be explained by the fact that while vinyl discs are played from the outside in, CDs are played from the inside out.

The 3-inch CD can hold up to about 20 minutes of music. Most releases had four popular songs. This format never really caught on with the buying public, and it has already been phased out by virtually all record manufacturers. It has been replaced by the "CD single," which is a full-sized CD containing just a few minutes of music. This seemingly wasteful format seems to have some popularity with teenagers who are only interested in top-40 hits.

The CD is a single-sided recording medium. There is no side 2. Actually, there is no technical reason why a two-sided CD would be impossible, but in practical terms, it would be a nightmare of production problems and quality control.

The label is printed on the CD's blank side. The CD is inserted in the player upside down, with the label up and the recorded side down. The CD player's laser assembly is mounted below the turntable and "looks" up at the disc. A simplified and greatly exaggerated cross-section of a CD is shown in Fig. 4-5.

Most of the CD's thickness is made up of the base material or substrate. The substrate is made of a polycarbonate substance. Encoded pits and flats are etched into the upper surface of the substrate. (This process will be described in more detail later in this chapter.) The etched surface of the substrate is then coated with a very thin layer of aluminum. This aluminum surface makes the pits and flats much more reflective. Finally, the aluminum surface is protected by a tough, clear acrylic coating.

It is this outer coating of acrylic that makes the CD so invulnerable to damage. Nothing can get to the actual recording surface to do any damage there. The acrylic coating is transparent, and minor damage to it is simply ignored by the laser when reading data from the disc. Severe scratches in the acrylic coating,

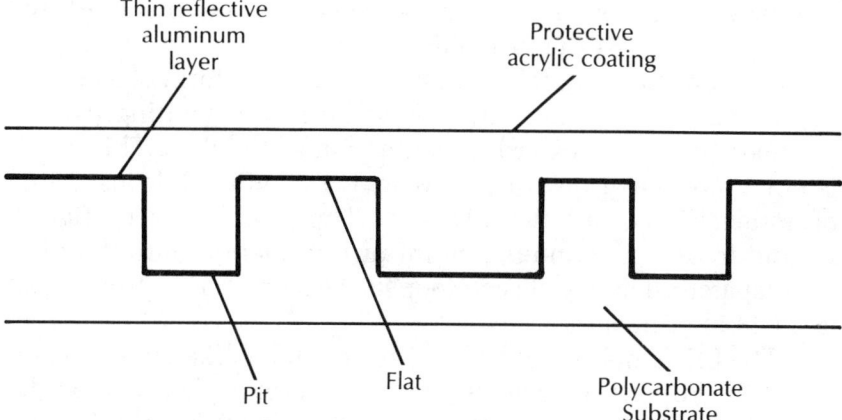

Figure 4-5 *This is a cross section of a CD.*

however, can result in losses of data because refraction effects can deflect light at certain points on the disc's surface. Fortunately, such severe scratches are fairly rare. Acrylic is naturally tough and scratch resistant, and the error correction circuitry in most CD players is quite good at coping with minor scratches. I've seen many CDs that were virtually covered with visible scratches that played just fine, sounding every bit as good as a mint-new disc. I have also seen a few CDs that were so badly scratched they couldn't be played at all. The nice thing is that such unplayable discs count for far less than 0.1% of all the CDs I've encountered. So, the CD pretty well deserves its reputation as being relatively indestructible.

To avoid distorting the optical data read from the disc, the CD must be perfectly flat. But no physical materials are perfect, and CDs, like vinyl LPs are subject to warpage, especially if exposed to strong heat over extended periods of time. A CD won't warp as easily or severely as a vinyl LP under similar conditions, but some warpage will still occur.

A slightly warped CD probably won't be a problem. The error correction circuitry of the CD player can handle the minor distortion of the read data. However, a badly warped CD may make it impossible for the laser to successfully recover the recorded data from the disc's surface. Tracking errors become increasingly likely with increased disc warpage.

In some, particularly severe cases, part of the disc could be warped enough to physically scrape the pickup device, potentially damaging it. A number of CD players are designed to guard against warped discs by including a special clamp that holds the disc tightly against the turntable platter.

The recorded digital data on a CD is in the form of a continuous spiral starting near the center hole and winding its way gradually outward toward the outer rim of the disc. This spiral is not a physical groove as on an analog vinyl LP. It is just a line of indentations and flat plateaus. The spiral is very tightly wound to fit a maximum amount of information onto the disc. The separation between each loop of the spiral varies from 0.833 to 3.054 micrometers.

The CD is not rotated at a constant speed like an analog record. To properly recover the recorded digital data it must be read at a constant rate. This means each lap of the spiral (complete path around the disc) must last the same time period. The laps near the center of the disc are obviously shorter than the laps near the disc's outer edges. This means that the revolution rate must change depending on where on the disc the laser is currently reading data. The turntable speed in a CD player varies from 500 to 200 rpm.

Along the length of the spiral are a series of oblong indentations, or pits, each exactly 0.11 micrometers deep. The individual pits are separated by level sections known as flats. In some technical literature, the flats may be called islands, or lands. There is no difference in meaning between these various terms.

During playback, a tiny laser beam is focused on the disc's surface and moved along the recorded spiral. At each point along the way, the laser light is reflected back to a photosensor. The photosensor sees more returned light when the laser beam is hitting a flat, than when it is focused on a pit. These light and dark pulses are converted by the photosensor into high and low electrical pulses, replicating the recorded digital data.

The length of the pits vary, depending on the data being recorded. The pit sizes vary in a stepwise rather than continuous fashion. (The pits on a videodisc vary continuously, or linearly, since the videodisc is an analog recording medium.) An individual pit may be as short as 0.833 micrometer or as long as 3.054 micrometer. Each discrete pit length has a specific meaning to the player's digital circuitry.

Roughly speaking, a pit on a CD can be considered a low, or a 0. A flat is a high, or a 1. However, there is not a bit-for-bit correspondence between the recorded data and the pits and flats on the disc. It might seem logical for each pit to represent a single 0 bit, and each flat to represent a single 1 bit. But this would be a very inefficient recording method. The available recording time would be considerably shortened. Any CD would be little more than a single if this recording method was used. Instead, the data is compressed and encoded into subcodes which assign specific multibit values to pits of specific lengths.

CD PLAYERS

A CD player, by necessity, is a very complex piece of machinery. Both the electronics and the mechanical portions of a CD player use advanced technology and high precision.

In operation, the CD is completely enclosed within the player. Most current CD players use a front-loading design, but there are some top-loading models available on the market.

In a top-loading machine, a lid on top of the unit is lifted and the CD is placed on the turntable platter. This system is somewhat similar in appearance to a traditional turntable for vinyl LPs. The lid must be closed for the playback circuits to function.

There are three basic approaches to front loading for CD players: drawer load, tray load, and vertical load.

To insert a CD into a player with a drawer loading system, the "eject" (sometimes called "open" or "load") button is pushed to open a sliding drawer in the front of the machine. Within this drawer are the turntable platter and the optical pickup. The CD is placed on top of the platter and the drawer is closed. The drawer must be closed in order for any of the player's other functions to operate.

A tray loading system is similar in operation, except the turntable platter and the optical pickup remain safely within the player. The tray is just a holder for the CD. When the tray is closed, a mechanism within the player positions the CD on the turntable platter for playing.

The tray system requires a little more in the way of mechanical gadgetry to automatically load the CD onto the platter. On the other hand, it offers somewhat greater protection to the platter and optical pickup mechanics than the drawer system. If nothing else, they aren't subjected to the wear and tear of being repeatedly moved back and forth. By far, the vast majority of CD players use either drawer or tray loading.

A very few CD players use vertical front loading. This works something like the top-loading system discussed earlier, except the disc is placed in the machine in a vertical (standing on its edge) position.

In all cases, the CD is loaded in the machine with the label side facing outward—the user can see the label as he's loading the CD into the player. The nonlabel side of the CD, which contains the actual recorded data faces into the machine. In all horizontally mounted platter systems, the data side of the CD is placed on the bottom. The optical pickup is underneath the platter, looking upward at the bottom of the spinning disc.

The reason for this is straightforward enough. The complete optical pickup assembly, with all of its motors, prisms, mirrors, and control circuitry is a fairly bulky item. Positioning it over the top of the disc platter would be physically awkward, and could make the player dangerously top heavy. Electronically, the optical pickup couldn't care less whether it was pointed up, down, or sideways.

A solid-state laser (also known as a laser diode) is used to read the recorded data from the surface of the CD. The laser diode itself is mounted on a movable arm. This arm is roughly equivalent to the tonearm on a traditional analog turntable. It's job is to position the laser at the appropriate place on the disc.

There are two types of pickup arms used in current CD players—the rotary arm and the sled, or slide arm. The rotary arm is very similar to the traditional tonearm on an analog turntable. The optical pickup is mounted on the free end of a pivoting arm. The free end of the arm containing the laser can swing back and forth in a shallow arc across the disc. Rotary arms are usually employed in top-loading and drawer loading CD players.

There are no physical grooves in a CD. On an analog turntable, the grooves in the record guide the stylus mounted at the end of the tone arm through the desired path. This isn't possible with a CD. A fairly complex automatic tracking system is used to guide the pickup across the recorded data in the desired sequence.

The other common type of pickup for CD players is the sliding arm, or sled. This system is somewhat similar to the lateral tonearms found on a few expensive analog turntables, and most analog cutting lathes.

The pickup assembly is mounted on a sled that moves back

and forth across the disc in a straight line. There is no pivoting base to this type of arm. Sled pickups are most commonly found in tray loading and vertical front-loading CD players.

Whichever type of arm is used, the laser beam is moved over the surface of the disc. The light is reflected back from the various encoded pits and flats. The flats will reflect more of the laser's light than the pits, so a photosensor can easily distinguish the difference between a pit and a flat.

The laser beam is not fired directly at the disc's data surface. First it passes through a complex series of lenses and prisms. The light pulses reflected back from the aluminum coating of the disc's pits and flats passes through a slightly different series of lenses and prisms to reach the photosensor, which converts the reflected light impulses into the appropriate electrical pulses. A high light level (from a flat) striking the photosensor causes it to generate a high electrical signal, while a dimmer light pulse (reflected from a pit) causes the photosensor to emit a low electrical signal.

Many current CD players are advertised as using three beams. Since only one pit or flat can be read at any given instant, this might seem pointlessly redundant. Actually only one of the three beams is used to actually read the data. The other two are used to help the player properly track the spiral of data etched into the disc's surface.

Two pairs of photosensors are used for this tracking technique, as illustrated in Fig. 4-6. If the beams are properly focused, as shown in Fig. 4-6A, each photosensor in either pair will receive an equal signal level. If mistracking occurs, it can go in either direction, too close or too far.

If the beams are focused too close, as illustrated in Fig. 4-6B, one of the photosensors in each pair will receive most or all of the reflected light, while the other photosensor will remain essentially dark. This imbalance will cause a proportionate error voltage to be generated. This error voltage will be fed to a set of servos, refocussing the beams further back.

If, on the other hand, the beams are focused too far, as shown in Fig. 4-6C, the process works in the opposite direction. The photosensors that received too much light in the too close situation now are too dark, and vice versa. The error voltage generated in this case is of the opposite polarity so the servos will move the beams in the opposite direction. This system will

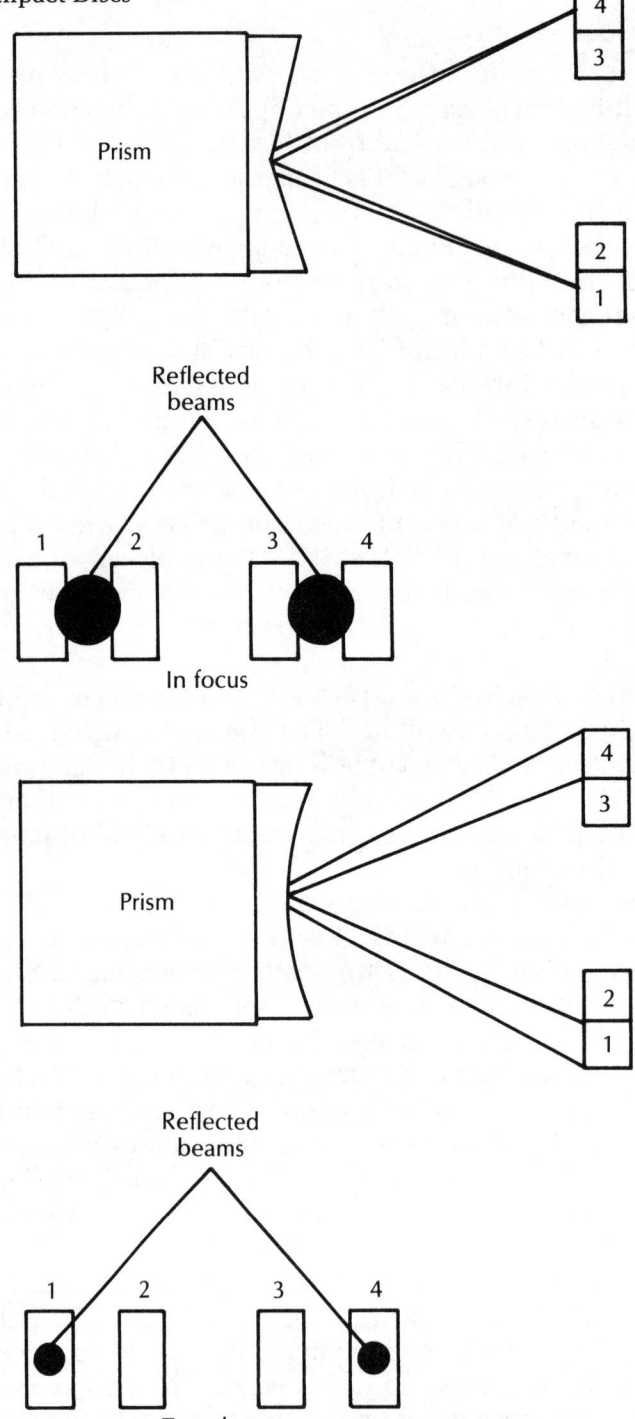

Figure 4-6 *Three beams are used for tracking the data spiral on the disc: a) on track; b) too close; and c) too far.*

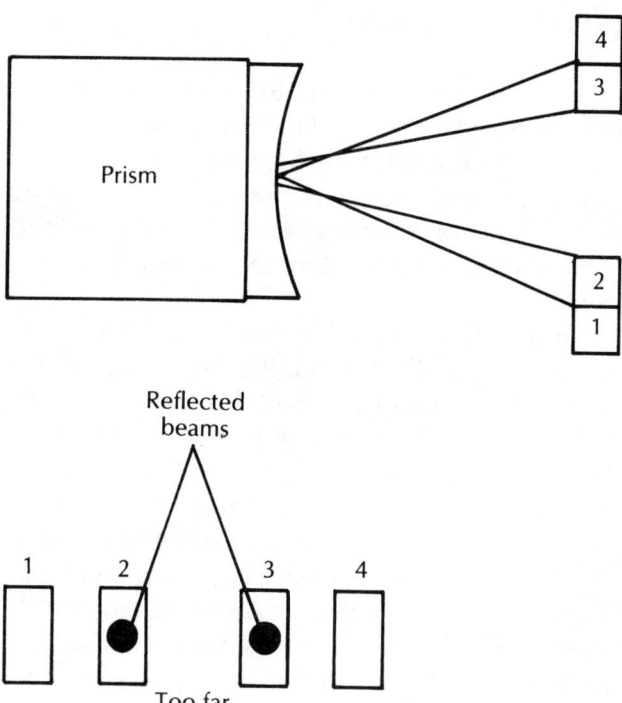

automatically keep the main laser beam perfectly on track to read the data correctly.

Few, if any, CD players use three separate laser diodes for this purpose. That would be too expensive and consume too much power. Instead, the beam from a single laser diode is passed through some special lenses, optically splitting it into three beams.

Unfortunately, the partial beams are inevitably weaker than the original unsplit beam. Usually this doesn't make much difference, but if the player is trying to read data from a disc with imperfectly formed pits, the split laser beam might not be sharp or strong enough. As a result, there is a higher chance of data read errors. In this case, the player's error correction circuitry would have to work that much harder.

The three-beam tracking system trades off an increase in data read errors for better tracking of the spiral of data. Other tracking systems are used in many CD players, but this is the most commonly used system.

COMPARING SPECIFICATIONS

A digital recording medium usually offers excellent specifications when compared with similar analog recording equipment. However, this may be a little misleading. Specifications that are critical in defining differences between analog equipment may be more or less irrelevant for digital devices; while other factors, unique to the digital format, may not be measured in standard specifications.

Some audiophiles aren't too impressed with digitally recorded sound. They hold that analog recordings are "warmer," or otherwise better sounding. An obvious example of inappropriate specifications is the wow and flutter specifications. (These terms are defined in Chapter 2.) Wow and flutter are of great importance for analog turntables and tape recorders, but such minor fluctuations in speed are utterly irrelevant in a digital recording. Wow and flutter will always be below measurable limits (or the minimum level indicated by the test equipment used) for any CD player or DAT recorder. There is no reason for manufacturers to include this specification for digital equipment, but it is almost always included. Presumably, this is because such specifications look terribly impressive and make appealing advertising.

A more crucial area of dispute in digital specifications is the distortion rating. This is normally given as total harmonic distortion (THD). Once again, all CD players and DAT recorders have excellent values in this area. THD ratings of 0.05% or even less are typical for digital audio equipment.

Does this mean the equipment is truly distortion free? Not necessarily. The THD measurements for digital audio components are made in the same way as for analog audio components. But digital distortion is quite different from analog distortion.

For example, standard THD ratings are weighted to account for the fact that analog equipment shows increasing levels of distortion with increasing signal levels. In a digital system, on the other hand, the distortion level will increase as the signal level decreases. Thus, the digital distortion may be underreported because of the way the measurements are weighted.

In fact, digital distortion may be more acoustically objectionable than analog distortion. In an analog sound system, the distortion increases during louder passages. The higher volume of the music itself can help mask some of the distortion effects. They are still there, but the human ear doesn't really notice them as much.

In a digital system, the highest distortion levels will show up during the quietest passages of the music, so the distortion effects will not be masked at all. They will tend to be quite apparent.

All in all, many currently used specifications don't appear to be particularly appropriate to digital sound equipment like CD players and DAT recorders. New measurement techniques and specification standards need to be devised.

There has been quite a bit of debate over whether all CD players sound alike. True, they usually have almost identical specifications from the cheapest models to the most expensive. But, as we've just seen, the specifications might not tell the whole story.

Many audiophiles insist they can, in fact, distinguish differences in the sound between various CD players. But the differences are certainly slight. Personally, I have heard a few minor differences between different CD players in simultaneous tests, where two units are compared side by side, frequently switching from one to the other and back again. But I was rather unimpressed by the differences. I doubt that I'd notice any difference if I listened to one player, then listened to a different one an hour later. Either one would suit me just fine in my home listening system. If I listen very, very closely, I can detect some tiny differences, but, in my opinion, they are totally on the "who cares?" level.

If you are a very critical listener, you should audition any audio equipment with your own ears before purchasing it. If you hear a difference that is significant to you, then your opinion is all that matters. You're the one who's going to be listening to the system, aren't you?

For most of us, when shopping for a CD player, there isn't much point in getting hung up on the specifications, or any slight differences in sound quality. Instead, you will want to confine your comparisons to three areas: price, durability and quality of construction, and features.

The price issue is pretty obvious. If you will get identical satisfaction from a $150 unit and a $500 unit, why throw away your money on the more expensive model? Price should not be the only consideration, however. The question is value, not literal cost.

This brings us to the question of durability and quality of construction. How well is the unit put together? Is it solid, or flimsily constructed and likely to fall apart in a few months? This is an especially important consideration with portables, or any equipment that is likely to be moved a lot or operated under adverse conditions of any type, but it is always a valid issue, even in the most protected applications.

The case is a good indicator of how well a piece of equipment is constructed. Does the case look and feel solid, or does it look like it will crack or shatter if it is dropped? (Certainly you want to avoid dropping any CD player, but, after all, accidents can happen.)

If possible, all other factors being equal, it is a good idea to opt for a machine with a die-cast metal housing. This is better and sturdier than thin sheet metal or plastic cases. Besides providing better durability, the die-cast metal case will offer better resistance to any airborne vibrations that could conceivably interfere with the laser's tracking.

Finally, there is the area of features. Think about what each special feature does and what it means to you, if anything. Some people are unduly impressed by a lot of snazzy buttons, whether they know what they're for or not. These people often waste their money paying for features they don't want or need, and won't use.

Different CD players offer a variety of programmability functions. Certain features you might use every day, while someone else will never have a use for them. Consider your own individual needs and temperment. Does the feature sound worthwhile to you?

Most CD players today, except for the lowest priced models seem to come with a remote control. This strikes me as one of the most pointless features. I can't figure out what it would be used for, except perhaps a pause button in case the phone rings while you're listening to a disc. My CD player does not have a remote control and I've never missed it. Other than the pause button, I have never had to get up to change a control during the playing of a disc. It is just as easy to do any programming at the front panel when inserting the disc.

As an extreme of the pointlessness of some remote control

functions, I've seen several models that have a button on the remote control for opening and closing the disc drawer. Why? No one is likely to want to change a disc from across the room. What are they going to do, toss a disc into the drawer from 20 feet away?

On the other hand, at least two audio equipment reviewers have written that they consider a remote control among the bare minimum features for an acceptable CD player. I'm not sure what they use their remote controls for, but they should look for a machine with a remote control. I'd prefer to not pay for something I consider totally useless.

No one can tell you which CD player (or any other type of audio equipment for that matter) is best for you. You need to make that decision for yourself. Unless you buy a real cheap model, that is obviously flimsy and poorly designed, you're not going to get stuck with a piece of junk when you buy a CD player. It's just a question of being a little bit satisfied or totally satisfied.

THE SPARS CODE

Most, though not all, commercial CD releases include a special three letter code somewhere on their packaging. This is known as the SPARS code, and it represents the types of recording used at various stages in preparing the master tape.

Each of the three letters in the SPARS code may be either an A for analog recording, or a D for digital recording at that particular stage. The first letter indicates whether the original recording was made on an analog recorder or a digital recording.

The second letter indicates the type of recording used for the mixdown stage. This is the stage where the various recorded tracks are mixed together into a suitable two-channel stereophonic recording.

The third letter is for the type of final master tape used. The actual data on the CD is derived directly from this master recording.

Since the SPARS code has three positions which can each take on either of two different values (A or D), there should be eight possible combinations. In practice, however, most of the possible combinations are never used.

It is theoretically possible to make a CD from an analog final master recording, but this is never done in practice. There would be no good reason to ever manufacture a CD directly from an analog master tape. It only makes sense to put the signal into digital form before the actual disc cutting stage. Therefore, all SPARS codes ending in A (AAA, ADA, DAA, and DDA) are never encountered in practice. I've never seen any commercial CD marked with any of these peculiar codes.

In effect, the third position of the SPARS code serves no particular purpose, since it always has the same value—D. Essentially, a two position code would work just as well.

Of the four remaining combinations, the DAD combination is rather ridiculous. Here we start out with a digital original recording, which is converted into analog form for the mixdown stage, then redigitized for CD production. There would never be any rational reason to work this way. Digital recording offers some of its most important advantages in the mixdown stage.

This leaves three practical SPARS code combinations—AAD, ADD, and DDD. All current CD releases that use the SPARS code use one of these three combinations.

An AAD CD starts out with an analog original recording. The mixdown process is also performed in the analog realm. This SPARS code is normally used with rereleases of old analog albums. Essentially, the original analog master tapes are simply digitized and then transferred to the CD.

If the SPARS code is ADD, we again start out with an analog original recording, but it is then converted into digital form to create the mixdown master. CDs of newer analog recordings are usually (but not always) of the ADD type. Some older rereleases also have an ADD SPARS code. Often, such discs are said to be "digitally remastered."

Finally, when a DDD SPARS code is used, you know the CD in question is 100% digital all the way. The original recording was made on a digital recorder. The mixdown master was also made in digital form, and of course, the final master tape was also a digital recording.

HOW A CD IS MADE

Interestingly enough, the process of manufacturing a CD quite closely parallels the process of manufacturing an analog vinyl LP. In manufacturing a traditional analog vinyl LP disc, once the master tape is complete, the recorded signal is fed to a special cutting lathe which cuts the modulated groove into a special mother disc. A mold is taken of this mother disc, and the actual final copies of the LP are pressed between these molds.

CDs are manufactured in a very similar manner. The final two-channel (stereophonic) digital master tape is transferred to a professional ¾-inch video tape recorder. The various subcode data fields are added to the tape at this stage. These subcode data fields are used for timing and indexing purposes in the CD player.

The digital playback signal from this tape recorder is fed into the cutting lathe to be etched onto the CD. This cutting lathe is rather similar in basic construction to a standard CD player, but a much more powerful laser is used.

The disc master is made of glass, coated with a layer of a photoresistive substance. As the master tape is played through the cutting lathe, the appropriate pits and flats are physically etched into the photoresistive layer by the laser beam. The etched glass master disc is then electroplated, usually with a thin layer of nickel to complete the master disc.

A mold is then made of the master disc. This mold is known as the father, or sometimes the matrix. In making the father mold, the original master disc is destroyed because the nickel plating is pulled off in the process. This thin plate is then used to mold several metallic mother discs. A mother disc is almost like a super-deluxe CD.

The final copies, or commercial CDs, cannot be made directly from the mother discs because they are in positive form. If a final CD was struck from the positive mother disc, the pattern of pits and flats would be reversed. That is, wherever there should be a pit, there would be a flat, and vice versa.

Another manufacturing stage is required. The mother discs are plated and then used to produce stampers. These stampers are then used to press the encoded data (pits and flats) into the substrate of the final CDs.

Once the substrate disc has been stamped, the center hole is punched out. The dimensions and positioning of this hole must be precisely correct, or the disc will be ruined. It will not be playable.

Next, the encoded surface of the substrate disc is coated with a thin layer of aluminum. This layer is extremely thin. It's thickness is specified at just 100 nanometers. The purpose of this aluminum layer is to make the pits and flats on the disc much more reflective, making the encoded data more easily readable by the CD player.

The reflective aluminum layer is applied to the substrate disc using a process known as vapor plating, which ensures that it will be smooth and evenly distributed over the entire surface of the substrate disc.

The next manufacturing stage applies a relatively thick protective layer of clear acrylic over the encoded surface of the disc. This is done with a spin coating machine. Ultraviolet light is used to bake the acrylic coating onto the disc. The purpose of this acrylic layer is, of course, to protect the encoded data (pits and flats) from potential physical damage.

Finally, the label is printed on the opposite (unencoded) side of the disc. The completed CD is then inspected for flaws, and packaged for sale.

❖ 5
The Basics of DAT

NOW WE FINALLY HAVE ENOUGH BACKGROUND INFORMATION AND we are ready to look more directly at DAT (digital audio tape).

When add-on PCM converters (used with VCRs) appeared on the market, everyone knew that something very exciting was starting. Some common models were the Sony F-1 and the DBX 700.

A PCM converter encodes the analog audio signals into a PCM (pulse code modulated) digital signal. This digital signal is then modified to resemble a standard analog video signal that can be recorded on a standard VCR. Video recording, like digital recording, demands a very large bandwidth, so the two systems have certain functional requirements in common.

The VCR itself does not need to be modified in any way in order to be used with a PCM converter for digital audio recording. But, while PCM converter/VCR combinations provided digital recordings with excellent fidelity, this was obviously not the ideal approach to digital audio recording on the consumer level. The merits of an all-in-one dedicated digital tape recorder were obvious. Combining the two units (the PCM converter and the VCR) was inconvenient, and quite intimidating to many people. For the most part, consumers were highly resistant to the whole idea.

Besides, the technology involved in the PCM converter/VCR combination was decidedly less than elegant. The PCM converter had to "disguise" the digitally encoded audio signals as video frames, complete with vertical and horizontal sync pulses to get the VCR to accept the data for recording.

There was also an inherent problem from the lack of worldwide standards. The digitally encoded audio data must be disguised as a video signal readable by the VCR, and different (and incompatible) video standards are used in various countries. The U.S. and Japan use the NTSC system, while Europe uses the PAL or SECAM formats. There are a number of extremely important differences between these various video systems. For example, the NTSC system uses 60 complete frames (full-screen images) per second, while in many European countries, the standard is a 50-Hz frame rate. If you attempted to feed a SECAM signal through a VCR or television receiver designed for NTSC signals, the result would be just so much meaningless and useless garbage. The electronic circuitry can't distinguish between the incompatible signal and a random noise signal.

This incompatability of video systems means that a tape made in Paris could not be played by someone in New York without special equipment. In order to create a workable consumer product, a worldwide standard was almost essential. All in all, the signs were unmistakable. The time certainly seemed ripe for a new product on the audio scene.

Digital recording of audio signals offers many significant sonic advantages over more traditional and familar analog recording techniques. Some of the most important improvements of digital audio recording include

- No surface noise or tape hiss,
- Very low distortion,
- Very wide and realistic dynamic range, and
- Very flat frequency response.

Of course, no practical technology is ever completely perfect. Digital recording of audio signals has certain inherent limitations and disadvantages of its own. The most significant of these disadvantages are

- High equipment cost,
- Quantization noise,
- Possibility of aliassing effects, and
- Impossibility of manual edits.

The high cost of digital recording is likely to come down as factories get into gear to supply the new technology.

Quantization noise and aliassing problems are not completely inescapable. Much can be done to eliminate quantization noise and aliassing or, at least, reduce them to negligible levels. It is more reasonable to consider quantization noise and aliassing as risks rather than true limitations of digital recording.

The editing problem, however, is almost certainly here to stay. Because of the way data is packed onto the tape in a digital recorder, manual cut and splice edits like those used with analog tapes, just won't work. When you cut the tape to make the edit, you'd cut across several of the diagonal tracks. Lining up tracks properly to make the splice would be impossible, because there is absolutely no visual indication on the tape as to where the tracks actually are. In addition, synchronization and control signals on the tape would almost certainly be totally confused by a mechanical splice.

There could also be major problems with feeding the spliced tape through the delicate and necessarily precise tape path. The thickness of the splicing tape is added to the thickness of the recording tape itself, and this could cause serious jamming problems. In short, mechanical editing just isn't an option when working with digital audio tapes.

On the other hand, electronic edits (performed by dubbing from one or more tape decks onto another recorder) are much more practical and efficient with digital recordings. There is no degradation of the signal quality, and numerous special effects can be added to the signal, if desired.

In making an electronic edit it is vitally important to match up the synchronization of the source machines and the destination recorder. If improper synchronization signals are used, it is impossible to correctly recover the recorded signal.

On the whole, the advantages of digital audio recordings seem to outweigh the disadvantages. This will be even more true when digital recording equipment prices come down.

Many professional recording studios have successfully employed digital recording techniques for years. Now, DAT recorders are being designed and marketed to bring these digital advantages into the reach of the general consumer and audiophile. For the time being, all DAT equipment is available only at top-of-the-line prices, but less expensive models are likely to appear in a few years. This is a very common pattern with electronic equipment. The early models of a new type of product are very expensive. In a sense, the first customers for the new product are paying for the research and development costs of the new product and retooling of the factories. Once a healthy market has been established for the new product, advances in the appropriate technology usually lead to a cheaper means of achieving the same ends. Some familar examples of this pattern are pocket calculators, VCRs, CD players, and personal computers. At first, all of these products were very expensive toys for the wealthy. Now they are reasonably priced, and many people wonder how they ever got along without them.

SETTING STANDARDS

A number of manufacturers were more than ready to set their research and development technicians right to work on creating a DAT recorder suitable for general consumer use. But they remembered the format wars (and lost profits) of several earlier products. The painful lessons learned from the quadrophonic fiasco of the 1970s (described in Chapter 1) were certainly not to be quickly forgotten. Several incompatable formats from different manufacturers succeeded in completely killing each other off in the marketplace. Even the phenomenally successful VCR has faced some extremely shaky times with the heated competition between the Beta and VHS formats, and later 8 mm and VHS-C.

The obvious solution for heading off such problems before they developed was for the various manufacturers to work together on setting industry-wide standards, rather than letting various separate manufacturers independently sink considerable amounts of time and money into the creation of multiple and incompatable solutions to the same basic design problems. Cooperation on standards had to come first, then the various companies could compete with specific products without needlessly confusing the consumer, and running down blind alleys of soon to be obsolete systems.

Format wars don't do anyone any good. Who wants to invest a large sum of money in a new type of tape recorder which will become obsolete in a few years, leaving the consumer unable to find either prerecorded or blank tapes? The consumer who backed the "wrong" format could well be left out in the cold with expensive but useless equipment.

To avoid such format wars and define worldwide standards in advance, about 84 different audio companies sent representatives to a worldwide DAT conference in Japan in 1985. Not surprisingly, Japanese manufacturers were definitely the most heavily represented at this conference, accounting for 60 of the 84 attending companies. The remaining 24 attending companies were based in North America, Europe, and elsewhere.

Thanks to this preliminary format setting conference, all DAT recorders will meet the same basic specifications and will offer full interchangeability between competing brands. A tape made on a DAT recorder made by company A can normally be played back without any problems on a DAT machine manufactured by company B, providing, of course, that machine A doesn't use an optional feature that is not supported by machine B. For example, all DAT recorders are required to support the standard 48-kHz sampling rate. The DAT standards also permit a slower speed using a 32-kHz sampling rate. This slow speed is optional. Manufacturers are in no way required to support this optional speed in all DAT equipment. So, a DAT tape recorded at 32 kHz may not be playable on all DAT machines. However, it will be fully compatible with all DAT recorders and players that do support the 32-kHz sampling rate. Naturally, all DAT tapes made at the standard 48-kHz sampling rate can be played back on any other DAT recorder or player. Some control codes for certain special features, such as some indexing tricks, may not be supported on all machines, but the tape will be playable.

This industry-wide standardization and compatability can only help the sales of all DAT equipment. The consumer doesn't need to guess which format will win out and which will end up obsolete and essentially useless in a few years. With one consistent standard, no one has to worry about getting stuck with equipment for a losing format. Nor is there any need for cautious customers to wait out a format war to find out which system to buy. There's just one standard format, so on that level, there is no competition to worry about.

Instead of a myriad of competing systems, the conference set standards for two basic (and incompatible) DAT systems S DAT and R-DAT.

The term S-DAT stands for stationary-head digital audio tape. S-DAT recorders are unlikely to ever become viable consumer-level products for various technical reasons.

An S-DAT stationary-head recorder has fixed-position record and playback heads, like a standard analog audio tape recorder. The tape must be pulled across the stationary head at a very high rate in order to provide enough room on the tape for sufficient data density. Digitally encoded data, as you should recall, always requires a very wide bandwidth, and that means a lot of tape space.

Because the tape must be fed past the stationary heads at such a fast speed, the motors and tape guides for an S-DAT recorder must be very precisely designed, which leads to very high equipment costs. Even a small error or imprecision in guiding the tape through its tape path could result in serious jamming problems. This could easily lead to broken or stretched (and therefore ruined) tapes, or even damage to the recording equipment itself.

It is not practical to whip the tape past a stationary head in an S-DAT recorder anywhere near as fast as the effective relative tape-to-head speed that is normally achieved in an R-DAT machine. Aside from the enormous mechanical problems, such an attempt would obviously result in extremely poor tape economy. An enormously long tape would be required to record even a short song.

Aside from the problem of incredibly poor tape economy, such high speeds would place extreme strain on the tape, the recorder's motors, and the tape guides. Tape breakage and jamming would almost certainly be a constant problem.

The only possible solution is for the S-DAT recorder to use a slower tape speed (with respect to the head), which inevitably means that there is significantly less room to store data on the tape. The recorded digital data must be much more tightly compressed, leaving far less margin for error. The tolerances for the circuitry in such a recorder must be extremely tight and precise. There is little or no room on the tape to store any error correction codes along with the actual recorded data representing the music.

Because of all of these inherent problems, it appears that S-DAT is not very viable as a consumer item; nor, is it likely to become any more viable any time in the foreseeable future. S-DAT recorders are intended primarily for use in professional recording studios. For the time being, the S-DAT standard has pretty much been dropped altogether.

The other standard DAT format is R-DAT, which stands for rotating-head digital audio tape. All consumer-level DAT equipment in the foreseeable future will be of the R-DAT type. In fact, the R-DAT format is so much more widespread that the "R" is often dropped from the acronym. Unless otherwise mentioned, it is usually safe to assume that any reference to DAT is probably talking about the R-DAT format.

For the purposes of this book, we won't be dealing with S-DAT equipment. Our primary concern here is with consumer-level equipment, which in this case definitely means R-DAT, not S-DAT. From here on, whenever we use the term DAT, we are referring to the R-DAT system.

A TYPICAL DAT RECORDER

Naturally, there will be many cosmetic and functional differences between DAT recorders made by different manufacturers. Different machines are quite likely to support different special features. Still, unless a designer really decides to go out of his way to be different just for the sake of being different, we can reasonably expect most consumer DAT recorders to more or less fit the following description.

At first glance, a DAT recorder looks very much like a CD player. On most DAT machines, the tape will be loaded into a slide-out drawer, similar to those used with most current CD players. Other types of loading bays are also possible, but for the time being the slide-out drawer appears to be the best and most practical choice in most cases.

The DAT recorder is a fairly substantial component, typically weighing about 25 pounds. A major reason for this relatively heavy weight is the fact that two separate and well-isolated power supplies will almost always be used for the analog and digital circuits. This means there will be two power transformers, which are quite heavy in themselves. The digital circuitry calls for a very well regulated, but relatively low-current power supply. The analog circuitry and mechanical portions of the DAT recorder, on the other hand, can accept greater fluctuations in the supply voltage, but consume far greater amounts of power. Since the power supply requirements for the digital and analog sections are so different, it makes sense to use two power supplies. A typical DAT recorder consumes a total of about 25 to 35 watts of power.

Another reason for the rather hefty weight of most DAT recorders is that most models (especially at the upper price range) will use a rigid die-cast nonmagnetic frame rather than a thin plastic or sheet metal frame.

As a general rule of thumb, a typical DAT recorder will probably have more controls than a typical CD player. These extra controls can be expected to include such things as address coding buttons and a record level control. These extra controls will be needed to accomodate the machine's various recording functions, which, of course, are not part of a CD player, which is a playback-only medium.

Despite the presence of these extra controls, DAT recorders will generally be quite easy to use overall—certainly a DAT recorder should be no more difficult to operate than a comparable CD player or analog cassette tape deck. In most cases, to play a DAT cassette, the user just has to insert the tape and press a button marked "play." I don't think it could get much simpler than that.

Display features of a typical DAT recorder include a meter to indicate the recorded signal's strength. This is very similar to a VU meter on an analog recorder. The signal strength meter is used to prevent overloading the tape or the recorder's circuitry with a signal level that is too strong. Other common DAT displays include various function indicators and time readouts. The function indicators display information about the machine's current operating mode.

On most DAT recorders, the time can be read in total time elapsed (since the beginning of the tape), program time, and time remaining on the tape. An automated sensor built into the recorder lets it know if a standard (2-hour) or thin (3-hour) tape is currently being used in the machine. A DAT recorder also automatically determines the correct playback sampling rate, and whether or not the "wide" playback mode is used on the tape being played. It will automatically set itself into the correct playback mode. (These various playback modes will be explained later in this chapter.) The user does not have to worry about which mode was used in recording the tape.

Some deluxe models of DAT recorders can be expected to have some extra readout features. A DAT's subcode area has about 4.6 times as much room as the subcode sections on a CD. This leaves plenty of room for the possible development of many new features and functions.

The DAT standards are quite extensive. Great efforts have been made to ensure compatibility between DAT recorders from different manufacturers. Still, we can probably expect some consumer confusion because manufacturers (and their advertising agencies) are free to use nonstandard names for many standardized features and functions. For example, the basically straightforward "next track" function could be jazzed up in ads with more "exciting" names such as MS (music sensor), AMS (automatic music sensor), TS (track search), and so forth. The only difference would be the legend printed on the appropriate button. The function is standardized. What the manufacturer calls it is not.

THE TAPE

Tape for a DAT recorder is enclosed in a plastic cassette housing, quite similar to that of an ordinary analog audio cassette. The tape in a DAT cassette is not quite as narrow as the tape in an ordinary analog audio cassette. The standard (analog) audio cassette uses tape of about ⅛ inch (3.2 mm) wide. DAT cassettes use tape that is a just a little wider. DAT tape is specified as 1/7 inch (3.81 mm) wide.

The tape thickness in a DAT cassette is specified as 13 microns, or about the same as the tape contained in a standard (analog) C-90 audio cassette. This allows a reasonable quantity of tape to fit into the compact cassette housing, without it being so thin that is excessively prone to jamming or breakage.

A standard DAT cassette contains enough tape for about 2 hours of recording time at the nominal (mandatory) running speed. A slower, secondary speed is also available on some DAT machines, increasing the total tape time to about 4 hours. In the wide playback mode (discussed later), the recording time of a standard DAT cassette is only about 80 minutes.

Some DAT cassettes will use a slightly thinner tape for up to 3 hours of recording time at the standard speed, and 6 hours if the slow speed is used. A small ID hole on the edge of the cassette housing will be used to tell the recorder how long the tape is and, in playback, whether a normal or wide track width was used in recording the tape. The recorder's circuitry automatically switches itself into the correct playback mode. The user never has to set any playback mode controls.

All recording times for DAT cassettes are for one direction only. Like video cassettes, DAT cassettes are recorded in one direction only. There is no side 2 on a DAT cassette. Once the tape has been played all the way through, it must be rewound back to the beginning of the tape to be played again. It cannot be turned over and played in the opposite direction like an analog audio cassette. In this respect, a DAT cassette is exactly like a video cassette. Like video cassettes, DAT cassettes are intentionally designed so that they cannot be inserted into the machine the wrong way. You cannot put a DAT cassette into a recorder backwards or upside down. This "idiot proofing" is considered necessary because of the problems that would arise if the recorder attempted to read the recorded data backwards.

The DAT cassette is quite compact and convenient. It measures just $2\frac{7}{8}$ inches wide, $2\frac{1}{8}$ inches deep, and less than $\frac{13}{32}$ inch thick. In metric values, the DAT cassette's dimensions are 73 mm × 54 mm × 10.5 mm. This is about half the size of an analog compact cassette, which measures 102.4 mm × 63 mm × 12 mm. A standard DAT cassette is illustrated in Fig. 5-1.

For most DAT recording purposes metal-particle tapes must be used. This kind of tape is more expensive than the ordinary ferric oxide tape commonly used in most analog tape recorders. Unfortunately, the magnetic particles in a ferric oxide tape simply aren't fine enough to accomodate the very high recording density requirements of the DAT system. The one exception to this rule is the wide playback mode. In this mode, the track

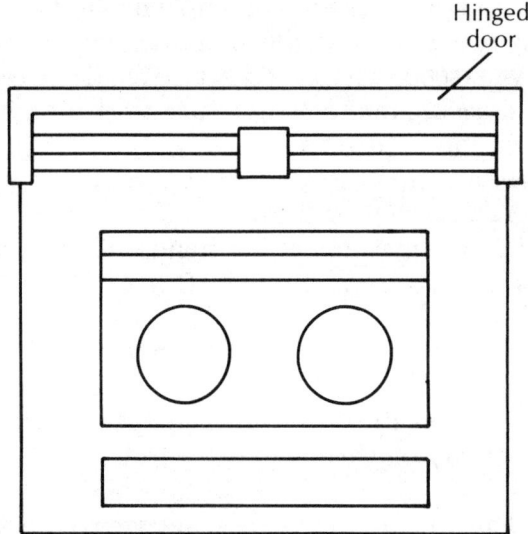

Figure 5-1 *This is a DAT cassette.*

width on the tape is made wider so the larger magnetic particles will work, and the less expensive ferric oxide tape may be used. Unfortunately, this widening of the data tracks in this special mode also translates to a substantial decrease in the tape's running time. Instead of the standard 2 hours of recording time, a standard DAT cassette can hold just 80 minutes of music in the wide playback mode. This is a playback-only mode to be used with prerecorded cassettes. Consumer-grade DAT recorders will not be designed to record in this mode.

There is currently some research being done into the possible use of other magnetic oxide materials, such as barium ferrite, in DAT cassettes. Generally speaking, the tape formulations used in DAT cassettes will be very similar to that used in 8-mm video cassettes.

THE HEAD DRUM

In a typical DAT recorder, a pair of record/playback heads are mounted on a rotating drum. The head drum and drive mechanisms in a DAT recorder are quite similar to those used in 8-mm video recorders (see Fig. 5-2).

The standard DAT specifications call for a head drum size of 30 mm (1.18 inches) in diameter. In normal operation, this head drum rotates at a speed of 2000 rpm. In the slow speed mode, the head drum rotation drops to 1000 rpm. For high speed search functions, the head drum rotation is sped up to 3000 rpm for fast forward, and a speed of 1000 rpm is used for rewind searches.

The effective tape speed in a DAT recorder is the combined relative velocity of both the head and the tape with respect to

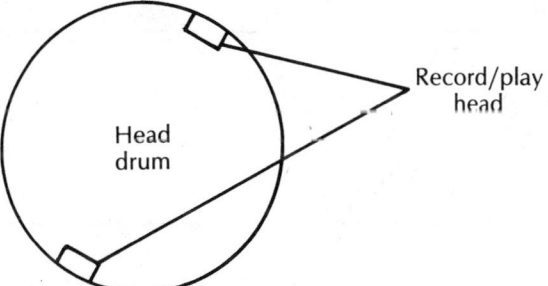

Figure 5-2 *Two heads are placed on opposite sides of the rotating head drum.*

one another. In a DAT recorder, the tape itself is actually moving past the head drum at a fairly slow rate. In the normal modes, the actual tape speed is a mere 8.15 mm/sec (about ⅓ IPS). This is extremely slow when compared to the tape speeds used in ordinary analog tape recorders. For example, a standard analog cassete tape recorder uses a tape speed of 1⅞ in./sec, or about 5.5 times as fast as the speed used in a DAT recorder; and cassette tape moves quite slowly by analog standards. A typical reel-to-reel tape recorder uses a tape speed of 7½ IPS, or 4 times the speed of the analog cassette tape, and 22.5 times faster than the actual speed of the DAT tape. The tape in a DAT recorder creeps along at a very, very slow rate. But, at the same time, the head is rapidly revolving in the opposite direction. As far as the heads are concerned, the tape is whizzing past at about 35 to 40 ft/sec.

The actual speed of the tape in a DAT recorder is about ⅙ the speed of the tape in an analog cassette recorder. Thanks to the rapidly revolving head drum, the effective speed (as seen by the heads) in a DAT recorder is about 66 times the speed used in an analog cassette machine. This permits an adequate data density on the tape.

The two heads mounted on the rotating drum are angled away from each other to avoid the possibility of crosstalk problems between adjacent tracks. One of the record/playback heads is set at an angle of +20 degrees while the second record/playback head has an angle of −20 degrees. This gives a total angle difference of 40 degrees between adjacent tracks. The angle of a recorder's heads is often referred to as *azimuth* in most technical literature. The basic track layout used in the DAT system is illustrated in Fig. 5-3.

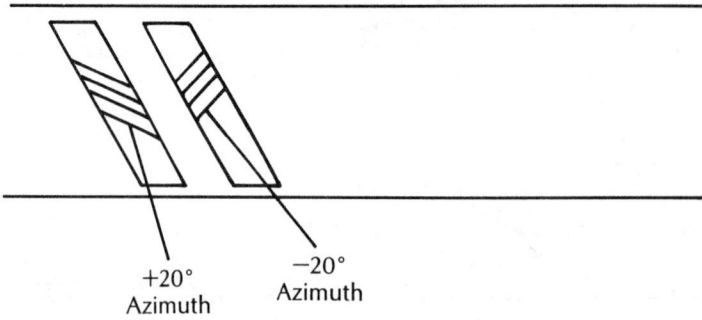

Figure 5-3 *This is the basic track layout used for DAT recordings.*

DAT recorders do not include erase heads. This may seem a curious and highly undesirable omission, but dedicated erase heads just aren't necessary for digital recorders. When a new signal is recorded it is strong enough to overpower and effectively remove any previously existing signal. This process is known as overwriting. Overwriting is used to avoid inadvertently erasing important control code data that the DAT recorder requires to operate correctly.

HOW DATA IS RECORDED

To prevent physical contamination of the tape surface, DAT tape is enclosed in a plastic housing or cassette. The user should never touch the actual tape. Remember, manual cut and splice editing is not possible with DAT cassettes, so there is no reason to ever touch the tape itself.

In an ordinary (analog) audio cassette, the tape never leaves the cassette housing during record or playback. This is possible because stationary record/play heads are used. However, with a rotating-head system, such as DAT, some of the tape must be pulled out of the cassette housing and threaded through a fairly complex tape path. This includes wrapping the tape partially around the revolving head drum. If this complex tape threading was done manually, the user would have to touch the tape in order to place it between the appropriate tape guides. Fingerprints and other contamination would be an inevitable problem.

In addition, the relative complexity of the required tape path could lead to serious problems if an error was ever made in threading the tape. For these reasons, a DAT recorder like a VCR always uses an automated mechanism to automatically pull some of the tape out of the cassette housing and thread it through the proper tape path. The user only has to insert the tape cassette into a well, and the machine takes care of the rest.

The tape in a DAT recorder is wrapped 90 degrees, or one-quarter of the way around the circumference of the head drum, as illustrated in Fig. 5-4. This is considerably less than the amount of tape wrap used in most standard VCRs.

There are two record/playback heads mounted on either side of the rotating head drum. The heads are separated by 180 degrees, directly opposite one another on the drum. Because of this arrangement, only one head is ever in actual contact with the tape at any given instant. In fact, there are short switching gaps

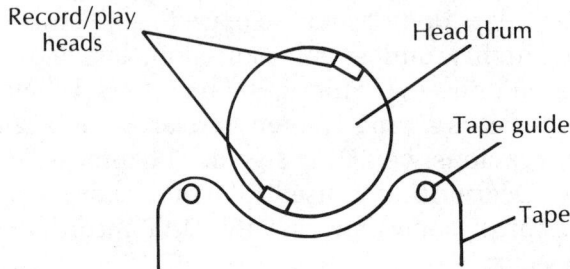

Figure 5-4 *The tape in a DAT recorder is wrapped 90 degrees around the circumference of the rotating head drum.*

when neither head is in contact with the tape. To compensate for these gaps, the signal data must be time compressed during recording to avoid the very serious possibility of losing portions of the signal whenever the head drum is positioned so that the tape's surface lies between the heads.

The DAT recorder's standard circuitry includes a special data buffer, which is a sort of temporary memory. During the recording process, the data buffer is constantly accepting incoming data from the signal to be recorded. Rather than releasing the output data in a continuous stream, the buffer spits out bursts of data in discrete blocks whenever either of the two heads is in contact with the tape. When the head drum is in a between-heads position, where neither head is in contact with the tape, the data buffer holds back its output. Data is fed to the record heads only at those times when they can actually transfer their magnetic field to the tape. Of course, only one of the two record heads can function at any given instant. In this way, none of the incoming signal data is lost during those brief times when both of the heads are out of contact with the tape.

On playback, this data buffering process is reversed, and the data buffer is now used for time expansion instead of time compression. It accepts the spaced bursts of data from the playback heads each time they come into contact with the tape, and releases the data in a continuous, time-corrected stream to the playback and digital-to-analog conversion circuitry of the recorder.

In order to successfully recover all of the recorded data on playback, the playback heads must make contact with the tape in the proper positions to match up with the recorded tracks. This positioning is very critical and exact. If the playback heads should happen to line up with the gaps between tracks (from the times when the head drum was in a between-heads position during recording), they obviously cannot possibly read the data stored on the tape. Obviously, if the playback heads tried to read the gaps between tracks, they'd pick up nothing at all. There is no signal recorded on the between-track gaps. To avoid such devastating misread problems, every DAT recorder includes a sophisticated automatic tracking correction system.

During recording, each time one of the heads comes into contact with the tape, it lays down one complete track, containing the burst of signal information from the data buffer. Each individual track is divided into five fields, as illustrated in Fig. 5-5.

Most of each individual track, of course, is made up of the actual PCM data field, which is located in the central portion of the track. This field is further subdivided as shown in Fig. 5-6. We'll discuss the specific divisions of the PCM data field shortly.

Flanking the block of PCM data are two smaller fields, labeled ATF (automatic track finding). These ATF fields are dedicated specifically to the task of supplying proper positioning information for the recorder's auto-tracking function. During the playback process, the head intentionally overscans the track it is

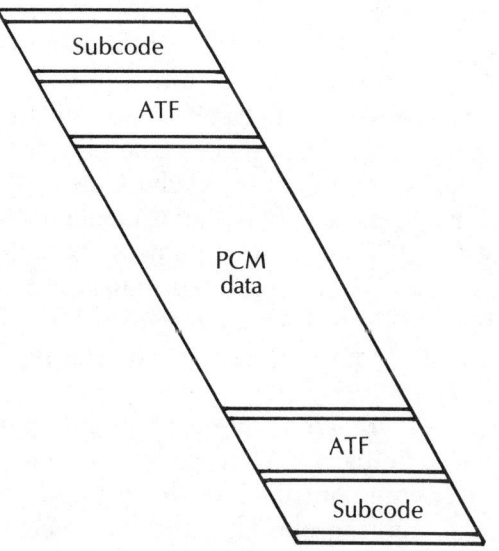

Figure 5-5 *Each individual track is divided into five fields.*

150 The Basics of DAT

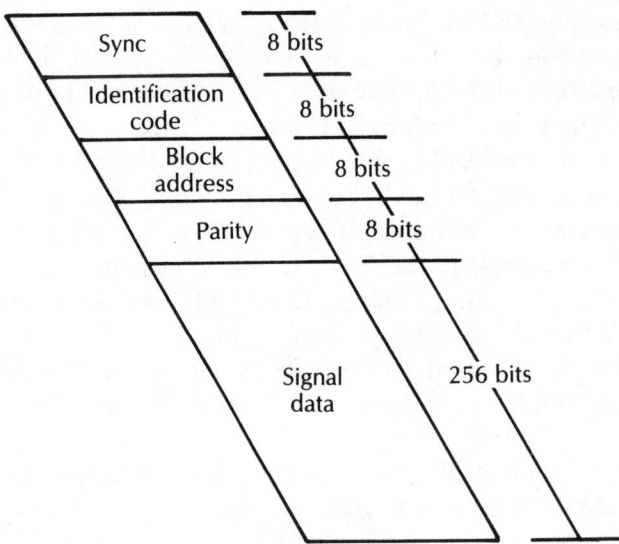

Figure 5-6 *The PCM data field includes the actual signal data and four smaller subfields.*

reading. That is, the playback head reads an area that is somewhat wider than the actual track laid down on the tape. Because of this deliberate overscanning, the head also picks up a little bit of the data from each of the adjacent tracks.

The overscan signal is sent to special circuitry that looks specifically at the ATF fields. The strength of the ATF signals on the two adjacent tracks is compared. If the head is correctly positioned, it should have equal amounts of overscan for the preceding track and the following track. If any inequality in the balance of the ATF overscan signals is sensed by the ATF detection circuitry, the recorder's tracking mechanism will automatically be readjusted to bring these two signal levels back into balance. When both of the ATF overscan signals are exactly equal in intensity, the recorder "knows" that the main track is properly positioned in the middle of the range read by the playback head. The playback head is now correctly tracking the data fields recorded on the tape.

Thanks to the automated self-correcting tracking feature permitted by the ATF fields, a DAT recorder does not have any need for a manual tracking control, like those found on virtually all VCRs. This is something else that the user never has to worry about. In effect, the DAT recorder automatically sets its own tracking control.

Continuing with our look at the various fields contained in each data track on the DAT tape, at the extreme ends of each track, beyond the ATF fields, are two small subcode fields. These fields are intended to contain miscellaneous nonmusical data, such as tape time, music selection control signals, indexing control signals, address control signals, search control signals, time control signals, and so forth. This subcode data does not contribute directly to the reproduction of the recorded music signal, but it is very useful in helping the recorder perform various "housekeeping" and convenience functions (which may or may not be supported on all commercial DAT recorders). Not all DAT recorders record all of the same subcode data. It depends on which optional functions the individual machine is designed for. If a given machine does not support a given function, it is quite reasonable that it doesn't record the appropriate subcode control signals for that particular function.

The subcode fields for DAT are relatively large. They are designed to hold up to 4.6 times as much data as the nonmusical data sections on a CD. Naturally, not all of this available subcode data space will be fully employed in all DAT recorders. This will be especially true for early units and later less expensive models. The designers of the DAT standard intentionally left more than enough data space in the subcode fields to cover every possible application they could think of, with plenty of extra space left over. This sufficient space leaves room for the system to grow and expand. The DAT standards are protected for any future time when someone dreams up new applications requiring more specialized subcode data.

The upper and lower ATF and subcode fields are generally duplicates of one another. Such duplication of data protects against possible data loss due to tape drop-outs or other minor problems. Not surprisingly, the greatest portion of each track is taken up by the actual PCM data field, in the center of the track. This relatively large data field consists of 288 bits of digital data, as illustrated in Fig. 5-6. Most of the PCM data field (256 of the 288 bits) is taken up with the data representing the actual audio signal that has been recorded.

The PCM data field also contains four 8-bit subfields that are used to perform various functions not directly involved with actually reproducing the recorded musical signal. For the most part, these four data subfields have pretty self-explanatory names:

- Sync,
- Identification code,
- Block address, and
- Parity.

While each of these names should be more or less clear by themselves, it is worthwhile to devote a few brief words of explanation to each of these data subfields.

Obviously, the sync portion of the data field is used to synchronize the playback circuitry. The playback circuitry uses the sync field data to know when to expect specific types of signal data. Without proper synchronization, the DAT recorder wouldn't be able to tell the difference between control data and the data used to represent the recorded analog signal. Obviously, correct reproduction of the recorded signal would be quite impossible under such circumstances. All 1s and 0s look exactly alike. The only difference is in the meaning assigned to them. Therefore, the various 1s and 0s must always be in exactly the right place at the right time, or the data will become hopelessly confused.

The identification code and block address subfields let the recorder know which block of data it is currently looking at, and where it belongs in the total reproduced signal. This permits the playback circuitry to reassemble the continuous signal from the various separate recorded tracks. The identification code and block address data may also be used in certain search functions to permit the recorder to automatically find a specific location along the length of the tape.

Finally, the parity subfield is used for spotting and correcting data errors in the recorded signal. If the data in the parity field does not match the parity value calculated with the recovered signal from the playback heads, the recorder knows some kind of error has been made in the recording/reproduction process. An incorrect parity code signals the DAT recorder's error correction circuitry that it's time to go to work.

In most cases, the error can be properly corrected and the

lost data estimated or replaced with few or no audible glitches appearing in the reproduced music.

Full stereophonic (two-channel) recording is assumed in the DAT specifications. Modern commercial recordings are almost always in stereo. (In fact, it might not be at all unreasonable to leave out the word "almost" from the preceding sentence.) Stereo is unquestionably the norm for modern audio applications. Curiously enough, stereo is even used for recordings of solo instruments. This may seem like overkill, but the fact is, the record companies cannot afford to maintain both monaural and stereophonic equipment.

Much of the public considers mono dreadfully old-fashioned and simply won't buy nonstereo recordings anymore. I suppose a case could be made that the increased ambient sound and reflections in a stereo recording of a solo instrument might improve the realism somewhat, but the effect will almost certainly be very subtle for most listeners. There is little reason to record a solo flute in stereo, but there is even less reason to avoid doing so if that is the way the equipment is already set up. At any rate, it was certainly a foregone conclusion that DAT would be a stereophonic recording medium. There was absolutely no reason to do otherwise.

In most of its operating modes, DAT uses two channels for stereophonic reproduction. It also has a quadrophonic mode with four independent channels. Since very few home audio systems are set up for quadrophonic sound reproduction, this mode will mainly be of interest to professional and semi-professional recording technicians using multitracking mixdown techniques. Four tracks of different instruments can be recorded and later mixed to the standard two stereo channels, with full control over the stereo imaging and relative levels. For the time being, we will concentrate only on the standard stereo (two-channel) DAT modes.

In a DAT recorder, the data for the two stereo channels is broken up in what may appear to be a very peculiar fashion, as illustrated in Fig. 5-7. Each track contains half the data for the left channel and half the data for the right channel. Each channel is broken up into odd and even blocks. The even data for the right channel is placed on the same track as the odd data for the left channel, while the odd data for the right channel and the even data for the left channel share the next recorded track.

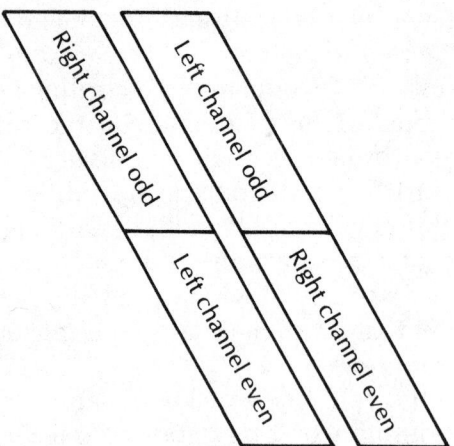

Figure 5-7 *Data for the stereo channels is broken up and interwoven to aid in error correction.*

It may look like the designers of the DAT system were being deliberately perverse and trying to overcomplicate matters. Actually, there is a very good and logical reason for this admittedly strange arrangement. By breaking up the channel data in this way, any little flaw on the tape (a dropout or a speck of dust or other contamination) won't destroy too much data for the recorder's error correction circuitry to recover. Data that is physically adjacent on the tape is not adjacent in the signal to be reproduced. Rather than one big hole in the data, which would be difficult for the recorder's error correction circuitry to compensate for, this arrangement is more likely to give us several little holes that can easily be filled using relatively simple interpolation and other error correction techniques.

It is estimated that under normal use, a typical DAT cassette can be expected to hold up to an average of 1000 recordings or playbacks before ordinary wear and tear on the tape causes too many dropouts for the error correction circuitry to successfully and reliably handle them.

This is clearly excellent durability for any tape format. After all, who would want to hear the same music more than 1000 times? Notice that this is just an average estimate of the DAT tape's lifespan. Some specific tapes won't last nearly this long, while others will last much longer.

As you can see, the methods of storing data on the tape in a DAT system has been very well thought out. This system was created from scratch specifically for DAT recorders. It was not simply borrowed and modified from some other pre-existing product. This indicates that the designers of the DAT standards had data integrity and maximum fidelity of audio recording as their primary goal. They did not want DAT to be a system of compromises. It was designed from the ground up to do the best possible job of signal reproduction.

SIGNAL SAMPLING

The DAT standard allows for several different sampling rates. The basic 48-kHz sampling rate is mandatory, and must be provided for in all consumer-grade DAT recorders and players. Notice that this standard sampling rate permits a slight improvement in the maximum frequency response over the 44.1-kHz standard employed with CDs. An optional secondary sampling rate (32 kHz) is also possible with some DAT recorders, permitting a slower recording speed.

The basic 48-kHz sampling rate is used for tapes made by the consumer himself. To discourage unauthorized duplication and pirating, commercially prerecorded DAT tapes employ a different sampling rate (see Chapter 6).

There are two different playback-only modes, called "normal" and "wide." The recording circuitry in a standard home DAT recorder does not support these playback-only modes. The "normal" mode uses the same sampling rate used for CDs—44.1 kHz. The same digital conversion circuitry can be used for both CD and DAT releases in the mastering plant. In addition, the same digital master tapes can be used for both DAT and CD releases of the same material. This has the advantage of reducing the manufacturer's production costs in supporting both formats (CD and DAT).

According to the DAT standards agreements, all home DAT units will be designed so that they are unable to duplicate this sampling rate in the record mode. Analog copies can be made (with the attending degradation of the signal fidelity), but not digital copies. An analog copy can be made by decoding the recorded digital data into analog form, then reconverting it back into digital form for rerecording. Obviously, this extra processing of the signal in the analog mode will result in some degree of loss in the fidelity of the recorded signal.

The copying issue is a very important one in the development of consumer DAT equipment, especially in the U.S. This hot issue will be discussed in some detail in Chapter 6.

The "normal" mode will be used for many standard prerecorded tapes in the DAT format, especially for longer works, such as symphonies and operas. This will also probably be the mode of choice for commercial releases aimed at audiophiles. It is more than likely that even more commercially prerecorded DAT recordings will probably be released in the "wide" mode.

The "wide" playback mode is designed specifically for tapes to be commercially duplicated at high speed. This form of duplication involves fast winding a blank tape together with a similarly fast winding master tape containing the signals to be duplicated. In this duplication process, the tapes may be wound at speeds up to 200 times the normal playback speed.

During the fast winding, the two tapes are put into contact with one another using specialized magnetic field focusing. As a result, the formerly blank tape now contains the same signal as the master tape.

This high-speed duplication process is not entirely perfect, of course. The duplicated signal level is significantly lower than for a tape duplicated at the normal playback speed. To compensate for this loss in signal level, the track width is increased to 1.5 times the track width used in the "normal" mode. This, naturally enough, is why it is called the "wide" mode.

Ordinarily, DAT recording demands the fine degree of resolution offered by pure metal particle tapes. Unfortunately, such tapes remain fairly expensive. Metal-particle tapes are used in some upper-end analog cassette tape decks, but for the most part, analog recording uses less expensive ferric oxide tapes. The required data density of DAT recordings is normally beyond the capabilities of ferric oxide tape, due to the relatively large magnetic particles. However, the wider track widths of the "wide" playback mode permit the use of less expensive ferric oxide tape. This could be a significant economic advantage for some commercially released recording.

A wider track obviously means that fewer tracks can fit onto a given length of tape. Overall recording density is lowered significantly in the "wide" mode. Because of this reduced recording density, a standard DAT cassette will have a shorter playing time in the "wide" mode. Where a standard DAT cassette recorded in the "normal" mode has a playing time of approximately 120 minutes, a standard DAT cassette recorded in the "wide" mode has a playing time of only about 80 minutes.

This may be a perfectly acceptable trade in some cases. On the commercial level it is much more economical to use high-speed duplication instead of real-time duplication. Mass production is faster and more efficient. Also, less expensive ferric oxide tape can be used for DAT recordings in this mode, instead of the more expensive metal particle tapes usually required for DAT recordings.

Because they are more economical to manufacture, "wide" mode prerecorded DAT cassettes will probably be significantly less expensive than prerecorded DAT cassettes using the "normal" playback mode.

Most commercial recordings to date have been considerably less than 80 minutes in length anyway. The longer playing time of the "normal" mode will surely be overkill for many commercial releases. Like the "normal" playback mode, the "wide" playback mode also uses the playback-only 44.1-kHz sampling rate to discourage unauthorized digital copying of commercially owned recordings.

In the DAT standards, there are two playback-only modes (or speeds), as outlined above, and two record/playback modes (or speeds). The standard record speed is mandatory on all DAT recorders and players. This mode uses a sampling rate of 48 kHz, as mentioned earlier in this chapter. Other than the difference in the digital sampling rate, the standard record/playback mode is technically similar to the "normal" playback-only mode.

Some, though not necessarily all, DAT recorders may also be equipped with an optional "slow" record/play speed. At this lower speed, tape economy can be increased, at some reduction in playback fidelity. The slow DAT speed uses a sampling rate of 32 kHz.

When using this slower speed, a standard DAT cassette can hold up to 4 hours of music or other audio programming. The trade-off here is that the maximum recordable frequency is lowered to about 15 kHz, instead of the full 20 kHz of the nominal human hearing response. This really isn't as big a drawback as it may seem at first glance. The frequency response of the human ear tends to decline with age, so many listeners cannot hear much about 15 kHz. Remember, many high-fidelity tape recorders are only rated for frequency responses up to 15 kHz or so.

An analog tape recorder can reproduce higher frequencies

than the specified maximum, but probably at an attenuated amplitude. For example, if an analog tape recorder has a frequency response rating of 50 Hz to 15 kHz (± 1 dB) it will almost certainly reproduce a 17-kHz signal, but the attenuation of this signal frequency is likely to be more than the specified 1 dB. The frequency response is less flat above 15 kHz.

In a digital tape recorder rated up to 15 kHz, however, the maximum frequency rating is more of an absolute limit. Higher frequencies cannot be reproduced by this system without introducing serious aliassing problems, so higher frequency components are deliberately filtered out of the signal before it is recorded.

For purposes of comparison, a frequency response graph for a typical analog recorder rated for a frequency response extending to 15 kHz is shown in Fig. 5-8, and a similar frequency response graph for a digital recorder rated to 15 kHz is illustrated in Fig. 5-9.

Normally, a resolution of 16 bits per sample is used in DAT recording. This gives 65,536 possible discrete amplitude levels that may be used for each sample. On the slow (32-kHz sampling rate) DAT speed, only 12 bits per sample are used. This reduces the required data density on the tape by cutting the data to be recorded by one-quarter, but it also limits the fineness of detail for the available amplitude steps for the recorded signal. Using

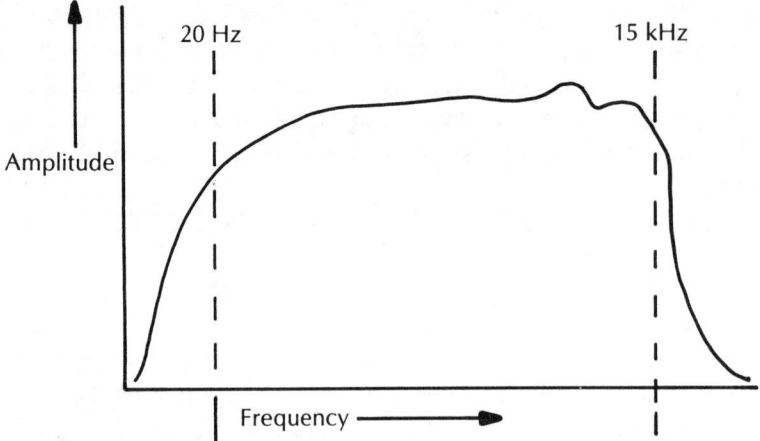

Figure 5-8 *This is a frequency response graph for a typical analog tape recorder rated to 15 kHz.*

Signal Sampling   159

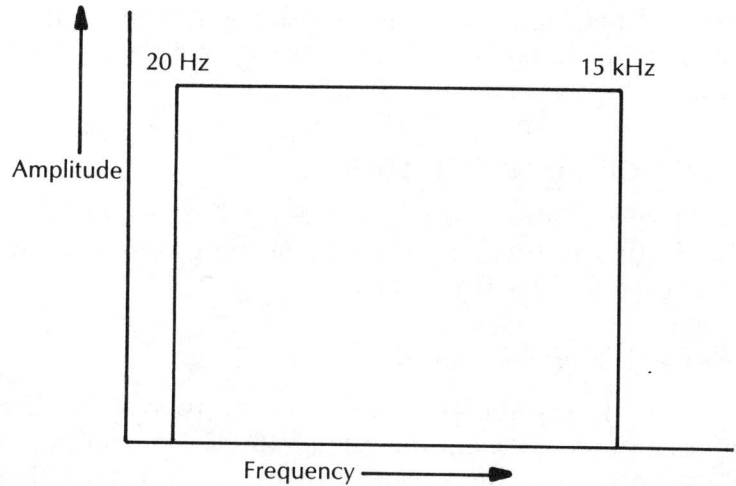

Figure 5-9 *This is a frequency response graph for a typical digital tape recorder rated to 15 kHz.*

12 bits per sample permits only 4,096 possible discrete amplitude steps. This is still a lot, but nowhere near 65,536.

The improved tape economy in this low-speed mode is traded for accuracy of fine detail in the recorded signal. In many noncritical applications—such as voice recordings and background music—this will be an entirely acceptable trade, but for high-fidelity, audiophile level recordings, the slow DAT speed may not be considered acceptable, even though the potential sound quality in this mode is significantly better than that of many low- to medium-priced analog sound systems.

I suspect that the average home listener probably won't really care too much about the very slight degradation of sound at this slow DAT speed. It is still a marked improvement over much of the analog audio equipment on the market today. It is likely that most home recording (except for true audiophiles, of course) will probably be done at this lower, more economical speed, to fit more music onto a standard DAT cassette.

The audio advantages of the higher recording speed are fairly subtle to an untrained ear. For the trained ear of the true audiophile, however, the 48-kHz sampling rate is almost a dream come true.

The DAT system has been carefully designed to fit the needs of almost everybody. The normal 48-kHz recording speed provides the wide frequency response and fidelity demanded by the serious listener, while the casual listener interested in background music will be delighted by the extremely long recording times available with the low (32-kHz) speed.

SUMMARY OF DAT SPECIFICATIONS

The DAT specifications provide for several speeds or operating modes. In this section, we will look at the technical specifications for each of these DAT modes.

The standard record/play speed

The standard record/play speed records two (stereophonic) channels and uses a sampling rate of 48 kHz. According to the DAT specifications, this speed must be supported in all DAT equipment. A DAT cassette recorded at the standard speed can be properly played back on any DAT recorder or player.

Since the sampling rate is 48 kHz, the frequency response of this mode covers the full 20-Hz to 20-kHz audible spectrum. The standardized DAT cassette can record up to 120 minutes (2 hours) in the standard mode.

The data resolution at the standard DAT speed is 16 bits. Linear quantization is used in this mode; that is, each discrete amplitude step is equal in size. A step from 0000 0000 0101 to 0000 0000 0110 represents exactly the same change in amplitude as a step from 1100 0110 1001 to 1100 0110 1010.

The head drum rotates at 2000 rpm, and the actual tape speed is 8.15 mm/s. This provides an effective writing speed in this mode of 3.133 m/s.

The recording line density at the standard speed is 61.0 kilobits/in., and the recording square density is 114 Megabits/sq. in. The subcode capacity is 273.1 kilobits/s. The transfer rate is 273.1 Megabits/s. This standard speed offers record and playback operations with excellent fidelity and very good tape economy.

Record/play option 1

The official DAT standards also provide specifications for three optional record/play modes. All three of these optional modes use a sampling rate of 32 kHz, limiting the upper frequency response to about 15 kHz.

Not all home DAT recorders will support all three of the

optional record/play modes. There possibly could be some compatability problems for tapes recorded using one of these options if different machines are used for recording and playback.

In Option 1, two (stereophonic) channels can be recorded and played back. Sixteen bits of resolution with linear quantization is still used, just as in the standard mode. The head drum still rotates at 2000 rpm, and the actual tape speed remains at 8.15 mm/s, so the writing speed for Option 1 is still 3.133 m/s.

The recording density for Option 1 is also the same as for the standard speed. That is, the recording line density is 61.0 kilobits/in., and the recording square density is 114 Megabits/sq in. The subcode capacity is 273.1 kilobits/s. The transfer rate is 273.1 Megabits/s, just as in the standard mode.

Like the standard speed, Option 1 offers record and playback operations with good fidelity and very good tape economy. The big difference between the standard speed and Option 1 is the reduced frequency response due to the lower sampling rate used in this mode.

The chief advantage of Option 1 is the somewhat lower recorded data density. In practice, this mode will probably be used only rarely. For the most part, this mode comes "free." Adding this option doesn't require any major additions to the recorder's circuitry, so it might as well be included, just in case.

Record/play option 2

The second optional record/play mode also uses a sampling rate of 32 kHz, and has a maximum frequency response of about 15 kHz. Once again, this mode supports two (stereophonic) channels.

The mechanical operations of the DAT recorder are slowed down for Option 2. In this mode, the head drum rotates at only 1000 rpm, instead of the usual 2000 rpm. In addition, the actual tape speed is cut in half to 4.075 mm/s. The writing speed in this mode is only 1.567 m/s.

This mode has only 12 bits of resolution. To compensate for the loss in resolution detail, nonlinear quantization is used in Option 2. This means the amplitude steps are not all the same size.

The recording line density for Option 2 remains at 61.0 kilobits/in., and the recording square density is still 114 Megabits/

sq in. The subcode capacity, however, is reduced to 136.5 kilobits/s and the transfer rate is 136.5 Megabits/s.

All of this adds up to a somewhat degraded fidelity of the recorded sound, but a very substantial increase in tape economy. Using Option 2, the recording time of a standard DAT cassette is doubled to 240 minutes (4 hours). The Option 2 mode has been referred to throughout this chapter as the "slow DAT speed."

Record/play option 3

The third record/play option is essentially a combination of the first two. The speeds and data rates are the same as in Option 1, but the fidelity is down to levels comparable with Option 2. That is because this mode supports four, instead of just two channels.

The sampling rate for Option 3 is 32 kHz, with a maximum frequency response of about 15 kHz. This mode uses 12 bits with nonlinear resolution, just like Option 2.

All of the other specifications for this mode are the same as those for Option 1. The head drum rotates at 2000 rpm, and the actual tape speed is 8.15 mm/s, so the writing speed for Option 3 is once more 3.133 m/s.

The recording density for Option 3 is also the same as for Option 1 and the standard speed. That is, the recording line density is 61.0 kilobits/in., and the recording square density is 114 Megabits/sq in. The subcode capacity is 273.1 kilobits/s. The transfer rate is 273.1 Megabits/s.

The chief advantage of Option 3 is the four-channel capability. This can be used for quadrophonic recordings, or sophisticated multitrack original recordings that can later be mixed down to stereo (two channels).

With the multiple track capabilities offered by this mode, different musical parts can be separately recorded then mixed down at appropriate levels. If, for example, a serious mistake is made in the lead guitar part, that track can be rerecorded without all of the other musicians having to play their parts over again too. If the drums are a bit too loud, that track can be attenuated somewhat in the mixdown process.

Since digital dubbing results in no degradation of sound, regardless of the number of tape generations used, the four tracks can be "bounced" in almost any way to give a virtually infinite number of available tracks.

The way this works can best be explained with an example.

The first three parts are each recorded onto their own individual tracks:

- Track 1, part A;
- Track 2, part B;
- Track 3, part C; and
- Track 4, blank.

Now, the first three tracks can be mixed together onto the fourth, free track. Then tracks 1 through 3 can be erased and reused. We can record two new parts on tracks 1 and 2:

- Track 1, part D;
- Track 2, part E;
- Track 3, blank; and
- Track 4, parts A, B, and C.

Then tracks 1, 2, and 4 can be mixed together onto track 3, then erased to record more new parts:

- Track 1, part F;
- Track 2, part G;
- Track 3, parts A, B, C, D, and E; and
- Track 4, blank.

This process of bouncing tracks back and forth can be extended for as many individual parts as you need. Unlike an analog tape recording system, you are not limited to just a few generations of tape before noise and distortion levels build up to an unacceptable point.

Even though the record/play Option 3 mode offers four independent channels, few DAT recorders will actually feature four output jacks for quadrophonic amplification. For the time being, a resurgence of public interest in quadrophonic sound reproduction seems rather unlikely. This mode will probably be considered the "multitrack mixdown" mode, rather than the "quad" mode.

The "normal" playback mode

The DAT standards also cover two playback-only modes used for commercially prerecorded cassettes. Consumer-level DAT equipment will not be capable of recording in these modes. The two DAT playback-only modes have already been covered earlier in this chapter, so this section may be considered as a review and summary of specifications.

The "normal" playback mode uses a sampling rate of 44.1 kHz. This is sufficient to allow a frequency response that covers the fully audible spectrum—from 20 Hz to 20 kHz. Two (stereophonic) channels are supported in this playback mode.

Other than the sampling rate, the specifications for the "Normal" playback-only mode are the same as for the standard record/play mode described earlier. The standardized DAT cassette can record up to 120 minutes (2 hours) in the "normal" playback mode. The data resolution in this mode is 16 bits with linear quantization.

The head drum rotates at 2000 rpm, and the actual tape speed is 8.15 mm/s. This provides a writing speed in this mode of 3.133 m/s.

The recording line density at the "normal" speed is 61.0 kilobits/in., and the recording square density is 114 Megabits/sq in. The subcode capacity is 273.1 kilobits/s. The transfer rate is 273.1 Megabits/s.

The "normal" mode offers only playback operations. No user recording operations are available in this mode. This mode features excellent fidelity and very good tape economy.

The "wide" playback mode

The "wide" playback mode is similar to the "normal" mode, except the required recording density on the tape is lowered significantly in this mode. Like the "normal" playback mode, the "wide" mode uses a sampling rate of 44.1 kHz. This is sufficient to allow a frequency response that covers the fully audible spectrum—from 20 Hz to 20 kHz. Two (stereophonic) channels are supported in this playback mode. The lower data density in this mode does not adversely affect the audio quality, only the tape economy.

Because of the lower recording density and a somewhat faster tape speed, the standardized DAT cassette can record only up 80 minutes (1 hour, 20 minutes) in the "wide" playback mode. The data resolution in this mode is 16 bits with linear quantization.

The head drum rotates at 2000 rpm in the "wide" playback mode. The actual tape speed is higher than normal in this mode. For the "wide" playback mode, a tape speed of 12.225 mm/s is used. This provides a writing speed of 3.129 m/s.

The recording line density in this mode is 61.1 kilobits/in., but the recording square density is only 76 Megabits/sq in. The subcode capacity is 273.1 kilobits/s. The transfer rate is 273.1 Megabits/s. The "wide" mode offers only playback operations with very good fidelity and fairly good tape economy.

Clearly, DAT is a very versatile system, capable of high-performance recording and playback for a very wide variety of potential audio applications.

6 ❖
Legal Issues of DAT

CONSUMER-LEVEL DAT RECORDERS ARE SLOWLY STARTING TO appear in the U.S. marketplace after several years of tantalizing promises and delays. DAT equipment has been available to consumers in Japan for some time now, but in the U.S., there have been several legal and political battles holding things up. In this chapter we will look at some of the legal problems that have blocked the U.S. sale of DAT recorders for so long.

THE GREAT COPYING DEBATE

One of the biggest selling points of DATs over CDs is that a CD is strictly a playback-only medium (at least so far—see Chapter 9). DAT allows the user to both play and record tapes. This recording capability is exciting for audio consumers, but rather worrisome for record manufacturers.

The recording industry's policing organization, the RIAA (Recording Industry Association of America), has long been concerned with (and perhaps overly preoccupied with) *pirating*, the illegal dubbing of copyrighted recordings. They claim that millions of dollars of potential sales are lost annually to illegal copies of recordings using existing analog equipment. They believe the problem will be greatly increased when digital recording capabilities are placed in the hands of the public.

Whenever a copy of a recording is made with an analog tape recorder, the sound is inevitably degraded somewhat. A legal, commercially marketed recording theoretically might be worth more than a pirated copy to serious listeners because of the better sound quality. But with digital recordings, commercial recordings lose this important advantage.

With a digital recorder, perfect copies can be made with no degradation of the signal. This is assuming direct digital-to-digital recording, of course. If the digitally recorded signal is converted into analog form and then redigitized for rerecording, this advantage is lost.

Perfect digital copies can be made because a digital recording consists of nothing more than a bunch of 1s and 0s. Tape hiss and distortion of the recorded signal have no noticeable effect (unless extremely severe) on the playback sound quality. When a digital tape is played, the data recovery circuitry automatically reshapes the modulation signal, and the reformed pulses can then be rerecorded without affecting the encoded data.

Digital copying can easily go through many, many generations (copies of copies) without any noticeable degradation of the recorded signal. A copy of a copy of a copy of a copy sounds every bit as good as the original master recording. Therefore, a pirated copy of a digital recording would not suffer in a quality comparison with a legal (commercial) copy of the same recording. A pirated copy can usually be sold for considerably less than a legal copy, because the pirate manufacturer doesn't bother with such expenses as royalties and original production expenses (studio time, mixing, etc.).

I don't think anyone would argue that pirated recordings are not a problem for the legitimate recording industry, but many believe the RIAA has greatly exaggerated the true extent of the problem. Their claim of millions of dollars of lost sales annually has been very seriously questioned in many circles.

First, a distinction must be made between the counterfeiter (or pirate) and the home recordist. The counterfeiter is in the business of making and selling illegal copies of recordings where the copyrights belong to others. The term "recording pirate" was coined to describe these counterfeiters. While counterfeiting is a legitimate concern, counterfeiters certainly don't account for "millions of dollars of lost sales annually." They are usually (though not always) small one- or two-person outfits making a relatively small number of counterfeit copies. The counterfeiter generally doesn't have access to very widespread distribution. They are also prone to severe cost cutting to maximize their profits. Most potential customers for a recording tend to be put of by the shoddy packaging used by most counterfeiters. Some more sophisticated counterfeiters try to duplicate the original (commercial) packaging of the recording. In this case, the goal is to fool the customer into thinking he is buying a legal copy. This type of counterfeiter is a definite minority, even among counterfeiters as a whole.

Few counterfeiters stay in business for long. Most quickly get caught. The smart ones only work their scam on a short-term basis, make a few quick bucks, then move on before they get caught.

Unfortunately, some degree of counterfeiting is always likely to be around, no matter what precautions are taken. If someone wants to do something badly enough (that is, if they think it looks profitable enough), they'll surely find some way around any safeguard. No technological or legal safeguard is ever 100% effective. What technology can do, other technology can undo. This is why the U.S. Treasury Department still occasionally arrests counterfeiters of currency, despite all attempts to make such illegal duplication as difficult as possible. The casual criminal is discouraged, but those who are determined enough will always find a way to get around the technological obstacles.

The RIAA has been insisting that all DAT recorders released in the U.S. include some sort of anticopying chip. This type of safeguard is highly unlikely to have any noticeable effect on the problem of counterfeiting. If one technician can figure out a way to put an anticopying chip into the recorder's circuitry, another technician can definitely figure out a way of taking it out, even if it means redesigning some of the circuitry. If enough potential profit is seen, this will be just a petty obstacle to the professional counterfeiter.

Moreover, many, if not most, large-scale counterfeiters presumably have some kind of inside connection with the recording industry and are likely to have access to studio equipment, which would obviously not include anticopying circuitry.

So the RIAA's campaign for an anticopying chip is intended to thwart illegal copying of commercial tapes by the home recordist. But do home recordists really represent that much of a threat to the recording industry? To support their claims, the RIAA points out that sales of blank cassettes are in the multimillion dollar range. What are all those blank tapes being used for?

A certain number of them are used for perfectly legal purposes that clearly have nothing whatsoever to do with copyright infringement or the legitimate interests of the recording industry. Some of the unquestionably legal recording applications include

- Answering machines,
- Dictation and note taking,
- Talking letters,
- Personal events and audio "scrapbooks" (such as a family reunion or baby's first words),
- Sound effects and recorded announcements (for theatres, store displays, etc.),
- Original music (garage bands, homemade demo tapes, electronic music composers, etc.).

Obviously, not all of those millions of blank cassette tapes are sold for the purposes of copying commercially recorded material. Still, the RIAA does have a point that all of these applications make up a definite minority of the home recording market. According to a number of surveys, most blank tapes are, in fact, used to make copies of preexisting recordings in the home.

But does all such copying really qualify as "pirating?" Does it necessarily mean a loss in sales for the recording industry? The existing evidence seems to indicate a strong "no" reply to these questions.

According to various surveys (and my own personal observation), the vast majority of home dubbing does not represent any loss in sales of commercial recordings at all. Far more than half of all the dubbing is done by someone who already owns a copy of the recording, and is making a copy for their own personal use. For example, many people make cassette copies of their records and CDs for use in the car, or a Walkman player. Party tapes and self-edited compilations are also popular among home recordists. Many people seem to enjoy making their own "best of" tapes. This is plainly a hobby activity, not pirating.

If anything, it appears that people who do a lot of home dubbing also tend to buy substantially more commercial recordings than the average audio consumer. These people are really into music. Discouraging these hobbyists could only hurt the recording industry. If they couldn't participate in their hobby by editing their own tapes, would they continue to buy so many commercial recordings? It seems many of these people would lose some interest under such circumstances. It's hard to imagine that cutting off their hobby would in any way lead them to buy more commercial recordings.

The DAT standard already includes some built-in protection against direct digital-to-digital recording of commercial releases, by using different sampling rates for the record/play and playback-only modes. Prerecorded DAT cassettes and CDs use a sampling rate of 44.1 kHz. DAT recorders can only record at sampling rates of 48 kHz or 32 kHz.

It is theoretically possible for an electronic whiz to get around this restriction and modify the DAT recorder's circuitry so that it will record at 44.1 kHz. But any technician who can do that, can certainly redesign the circuitry to bypass any other anticopying scheme. No one is ever going to be able to devise a technological method of stopping such technical geniuses. What technology can do, technology can undo. No technology is perfect.

In my opinion (shared by many people in the audio industry and many audiophiles), the RIAA is somewhat paranoid about something which is basically a nonissue in the real world. Nevertheless, many record manufacturers and the RIAA fear that unrestricted digital recording could lead to a substantial decrease in the sale of commercial recordings. To protect themselves, they have been trying desperately to get Congress to pass a law requiring all DAT recorders sold in the U.S. to have some sort of additional anticopying protection built in.

Of course, manufacturers could market DAT equipment without an anticopying chip before such a law is passed. It wouldn't be illegal until (and if) the law passes. But DAT manufacturers have been understandably cautious. It would be terribly expensive and inconvenient to start marketing DAT machines and then have to redesign all models to suit a new law. Any back inventory probably could not be sold. And how can any manufacturer be sure of conforming to a law that doesn't even exist yet?

In addition, while it is not yet a crime to sell an "unprotected" DAT recorder, that doesn't mean the RIAA, or one or more record manufacturers couldn't sue. In the 1970s, Sony (the manufacturer of Betamax VCRs) was similarly sued by the movie industry. Sony eventually won their case in court, but the legal battles were long and expensive. Clearly, no audio manufacturer wants to set themselves up as a test case in this type of conflict.

Who can blame them for cautiously holding back until the "ground rules" are officially defined?

So some sort of anticopying device seems inevitable for DAT recorders in the U.S. Even if Congress doesn't give such a system the force of law, almost all manufacturers will probably willingly comply with an industry standard for copy protection if one could be officially and universally agreed on. Most manufacturers of DAT equipment will probably include the anticopying protection to protect themselves from lawsuits from the recording industry.

THE NOTCH THAT DIDN'T WORK

If the recording industry is going to demand all DAT recorders be equipped with copy protection, they need to come up with some appropriate technology to do the job. This has proven to be easier said than done.

The first serious contender for the official copy protection standard was a notch system devised by the CBS Technology Center. CBS claimed this system could be supported by a single IC chip installed in every DAT recorder. They predicted a per unit cost for these copy protection chips to be approximately $1, in industrial quantities.

Almost as soon as this system was announced, protests against it began. Many technicians doubted the necessary circuitry could be successfully contained in just a single chip. There was even greater doubt that the system could be implemented into consumer recording equipment at a cost of just $1 or so.

Ultimately, these complaints were relatively trivial. There were far more serious charges against CBS's proposed copy protection system. Most of the protests against the CBS notch system centered on the serious doubts that it would work as claimed.

The CBS Technology Center was closed down as an economy measure in 1986, but the Records Division of CBS continued to back the system and led a strong push for Congress to pass a law requiring that all U.S. DAT recorders include the chip, even though the chip itself didn't exist yet.

The idea behind CBS's anticopy scheme was for protected recordings (or software, as it is sometimes called) to include a narrow notch at a specific frequency. If the anticopy chip sensed a notch at this frequency range, it would disable the DAT record circuitry.

The whole idea seemed more than a little unlikely right from the start. Most of the audio industry was dedicated to making the frequency response of the entire audio reproduction chain as flat as possible. Now, commercial recordings would be deliberately released with a decidedly nonflat frequency response.

CBS insisted that the notch would be so narrow that the effects would be inaudible. For a long time they were pretty closed-mouthed about the details of the notch filter design they proposed using, and especially the width of the notch and its roll-off characteristics—that is, how sharply the filtered signal was cut off. The system called for a pretty deep (strongly attenuated) notch, so inevitably, some adjacent frequencies must be affected.

No practical filter is ever perfect. With the notch (or band reject) filter, like the one in the CBS copy protection system, any frequency component within the specified band would be removed from the signal, while all frequency components outside the reject band would be completely unaffected. A real world filter doesn't work quite so neatly.

The notch is defined by two cutoff frequencies—the upper cutoff frequency and the lower cutoff frequency. For purposes of illustration, let's assume we have a notch filter with an upper cutoff frequency of 2000 Hz, and a lower cutoff frequency of 1500 Hz. A 500-Hz band is notched out of the signal. In a theoretically ideal notch filter, a frequency of 1501 Hz would be completely rejected, while a frequency of 1499 Hz would be completely passed with no attenuation. A frequency response graph for this impossibly perfect notch filter is shown in Fig. 6-1.

In a practical filter, the cutoff is not instantaneous. Instead, there is a gradual cutoff slope, as illustrated in the practical notch filter frequency response graph of Fig. 6-2. The cutoff slope indicates a gradual increase of attenuation with changes in signal frequency. The nominal cutoff frequencies are the points where the signal has dropped 6 dB from its nominal unattenuated level. Nearby frequencies are also attenuated to a greater or lesser extent.

The notch also won't be completely perfect. Some of the frequency components, even in the center of the notch band, will

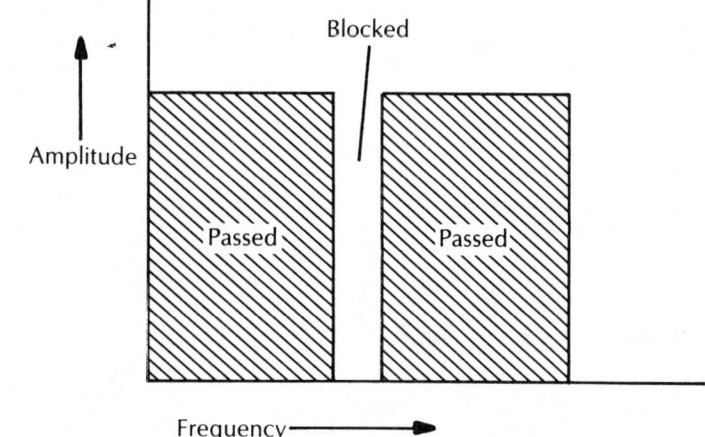

Figure 6-1 *This is a frequency response graph for an ideal notch filter.*

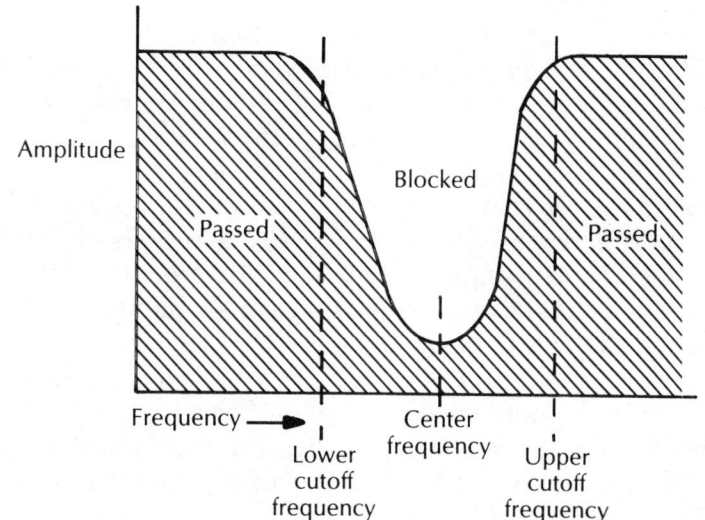

Figure 6-2 *This is frequency response graph for a practical notch filter.*

leak through the filter. The filter's attenuation is not infinite. A strong signal within the rejected notch will not be reduced to an amplitude of zero.

Of course, better quality filter circuits will exhibit deeper attenuation notches and steeper cutoff slopes, but neither feature will ever be infinite in any real-world filter. It is not possible to design a practical filter that can completely notch out a single frequency without also affecting neighboring frequencies.

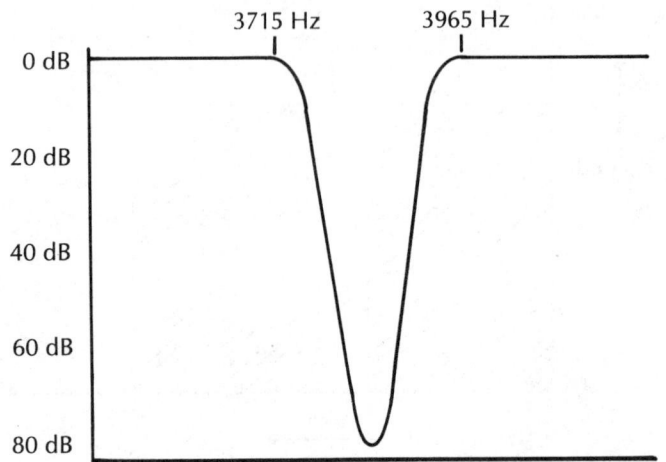

Figure 6-3 *The CBS Technology Center recommended using a notch filter in the audible frequency range for anticopy protection.*

The notch filter in the CBS system was centered around 3,838 Hz, which is clearly well within the audible frequency range. Generally speaking, musical notes are lower in frequency than this, but there are occasional exceptions.

Eventually, it was revealed that the notch filter had a bandwidth of 250 Hz, as illustrated in Fig. 6-3. Frequencies from about 3,715 Hz to approximately 3,965 Hz are attenuated by at least 3 dB. A difference of 3 dB represents a halving of the perceived volume of a sound.

Notes A7 and B7 (in the highest octave on a piano keyboard) both fall into the notch region. Note A7 has a frequency of 3,720 Hz and B7 has a frequency of 3,951. Surely the effects of the notch would be very audible if any musical notes this high were used in the recording.

To make matters even worse, the protesters pointed out that the notch filter would unquestionably have an audible effect on the harmonic content of many lower-frequency notes. Musical tones (in fact, most sounds) do not consist of just a single frequency component. They include many higher frequencies called harmonics at exact multiples of the base (or fundamental) frequency. The harmonic content of a sound has a major effect on its tonal quality. Therefore, it would seem highly likely that the following notes would probably suffer some noticeable sonic degradation from the CBS notch filter (the notes first on the list would suffer the greatest, most noticeable effect; the notes toward the end of the list would suffer lesser but still real effects): A6, B6, A5, B5, A4, B4, D5, E5, A3, B3, A2, B2, C4, D4, A1, B1, A0, and B0.

As you can see, music in just about every audible octave would be affected to some extent by this filtering. Yet, CBS continued to claim that their notch system would not have any noticeable effect on the sound of a protected recording. Then they seemingly contradicted themselves by saying that the recording company would always have the option of not using the notch filter at all, or of switching it off during those moments when it might degrade the musical content of the recording.

Opponents argued that if there was any signal content in the notched frequency range to filter out, its removal would always have an audible effect. CBS, the RIAA, and other record manufacturers strongly lobbied Congress to enact a law requiring that the CBS anticopy chip be included in all DAT recorders sold in the U.S. This was despite the fact that the chip needed to implement the system didn't even exist yet. So far, test setups have been made up of discrete components, so they are large, bulky, and expensive, and therefore not feasible for use in consumer equipment. Fortunately, Congress decided to wait until the National Bureau of Standards (NBS) checked the system out.

In the meantime, CBS Records announced that they would immediately start including the protective notch in their products. A number of audiophiles started having second thoughts about buying CDs on the CBS label. They felt that the company was marketing a deliberately inferior product.

It is not clear whether CBS ever released any copy protected CDs using their notch system. They just publicly announced their intent to do so; that doesn't necessarily mean they followed through. Perhaps CBS intended to use the confusion as a test of the audible effects. If no one could tell which discs were copy protected and which weren't, it would suggest that the system was inaudible, as CBS claimed.

The results of the NBS tests of the notch filter anticopying system were certainly disspiriting to the CBS technicians. The NBS report stated that in many cases the notch filter definitely caused audible degradation of the recorded sound.

Moreover, the reliability of the anticopying system was, in a word, lousy. It didn't always work. Sometimes a recording with the notch would not deactivate the DAT recorder's record circuitry. In other cases, some nonprotected signals caused false triggering of the anticopy mechanism. The system didn't always do what it was supposed to do, and it often did what it wasn't supposed to do. The conclusion was clear—the CBS system was a total failure.

But the RIAA and the record manufacturers continued to insist that some form of copy protection be included in DAT recorders. Manufacturers of DAT equipment and potential customers for such equipment were again stuck playing a waiting game. Somewhat ironically, shortly after the CBS notch system was squashed, CBS Records was bought out by Sony, one of the leading corporations in the development of DAT recorders.

THE SCMS COMPROMISE

A later system, developed by Phillips in late 1989 and early 1990, seems likely as an acceptable compromise for both the DAT manufacturers and the recording industry. Representatives from both sides have expressed at least grudging acceptance of this new system. Even if Congress does not pass a law making it mandatory, it is very unlikely that many DAT recorders intended for sale in the U.S. will not include it.

The Phillips system is called the Serial Copy Management System (SCMS). Unlike the CBS notch filter, SCMS works entirely within the digital subcode data, so it can have no possible effect on the reproduced sound.

The SCMS copy protection system will also have no effect on recording from analog sources. You can play a DAT cassette and convert the signal into analog form, then redigitize and rerecord it with a second DAT deck, with no problems from SCMS.

The digital codes on each DAT cassette (and each CD) will contain a single bit indicating whether the recording is copyrighted material. A 1 for this bit indicates yes, and a 0 means no. DAT recorders with SCMS will recognize this copyright bit whenever a copy of a digital signal is made. If the copyright bit is a "no," the recording will be made in the normal way. If, on the other hand, the copyright bit is a "yes," SCMS will allow the copy to be made, but in the process it will change another bit on the tape being made from a 0 to a 1. This bit is the actual copy-protect bit.

Now, if you try to make an additional copy of this copy, the DAT recorder's circuitry will detect the copy-protect bit, and refuse to let the record circuitry function. A home user can make a back-up copy, or edit his own "compilation" tapes from any recording in his collection. But he can't make an infinite number of digital copies of SCMS protected recordings. This will discourage the real pirates who make bootleg copies of copyrighted recordings for the purpose of selling them, cheating the legitimate record companies out of sales.

Of course, it is possible to copy a prerecorded tape or CD over and over, as long as the original is used as the source each time. This is awkward and time consuming, and would severely cut into the profitability of bootleg recordings.

All in all, SCMS seems like a fair and reasonable compromise to the copying problem. Legitimate users (the majority) are not penalized, but much of the criminal element (the minority) is somewhat stymied. Of course, a talented technician with a dishonest bent could probably figure out a way to defeat the SCMS system, but this is equally true of any conceivable anti-copying system. Whatever technology can do, technology can undo, if you're determined enough. But redesigning the circuitry of DAT recorders to get around SCMS would be rather inconvenient and expensive, and would inevitably eat into the potential profits of making illegal recordings in the first place. Presumably, SCMS will be discouraging enough that most recording pirates will decide its not worth their while.

A fringe benefit of SCMS is that it could be used for any digital medium. If at some future date we have digital radio transmissions, the home listener could record a program (like time-shifting with a VCR), but could not make copies. The same thing could work if a practical and commercially acceptable digital television transmission system is ever devised.

The best feature of SCMS is that it is virtually invisible to the honest user. Most consumers probably won't even know its there. SCMS does not affect the recorded sound in any way. In fact, I expect service centers will get a number of complaints from people who try to make copies of copies and can't figure out why it won't work. Many of these people will surely assume that there is something wrong with their machine. Perhaps the biggest disadvantage of the SCMS anticopying system is that service technicians will have to waste a little time explaining to their customers that nothing is wrong, that's the way DAT recorders are supposed to work.

THE GRAY MARKET

DAT equipment is just beginning to appear on U.S. dealers' shelves as this is being written in 1990. DAT recorders have been available in Japan and several other countries since shortly after the standards setting conference. Things have been held up here in the U.S. because of the legal tangles surrounding the copying problem discussed in the last few pages. The SCMS compromise seems to have opened the gate for DAT to finally reach American consumers.

A number of pre-SCMS DAT recorders have already been sold in the U.S., however. These sales are referred to as the "gray market." They aren't strictly illegal, like the "black market," but they're not entirely legitimate either.

Here's how the gray market works. An American dealer goes to Europe or Japan and buys a number of DAT recorders. Then he brings them back to the U.S. and sells them from his own store.

Gray market DAT equipment was eagerly welcomed by some audiophiles who just couldn't wait until DAT recorders overcame all the legal hurdles and were officially available. But there were some hidden costs attached to shopping on the gray market.

Of course, gray market prices were higher because the merchandise didn't go through the ordinary wholesale chain. The gray market dealer had to buy his stock retail, rather than wholesale; not to mention the costs of the overseas trips, shipping, and customs duties.

In addition, since the gray market equipment was not authorized for U.S. sales, manufacturers were not legally obligated to honor any sort of warranties or service agreements. Of course, the official U.S. distributors did not like the gray market dealers. If, for example, Sony made it a policy to honor warranties for gray market goods, the official Sony distributors would see that as unfair competition, since they had distribution agreements for directly marketing Sony products. If they resented the grey market competition enough, the legitimate dealers might well decide to stop carrying Sony products, and concentrate on Pioneer instead. If there was any kind of exclusive dealership agreement, the distributor could easily sue Sony for supporting sales through alternate, gray market channels. Needless to say, Sony (and other manufacturers, of course) wants to keep its customers happy, so they almost always flatly refuse to honor any gray market warranties.

Gray market sales of DAT recorders helped fan the flames of the great anticopying debate on all sides. If gray market sales managed to reach an extensive level, the whole anticopying issue would have been rendered a moot point.

7 ❖
DAT Equipment

IN THIS CHAPTER WE WILL LOOK AT SOME SPECIFIC PIECES OF DAT equipment now (or soon to be) on the market. There aren't very many DAT machines available yet, but that is likely to change in the near future.

Currently, because a cloud of legal questions still hangs over the DAT recording format, several manufacturers seem to be playing it safe. For the time being, automotive playback-only decks are somewhat more available than home DAT recorders. This too is likely to be a temporary situation. In this chapter we will look at both automotive and home DAT recorders.

AUTOMOTIVE DAT DECKS

Most of the first wave of legitimate U.S. DAT equipment seems to be in the realm of automotive DAT decks. For playback-only decks, the troublesome question of copy protection is irrelevant. However, for these DAT units to sell well, they must be supported by a strong selection of prerecorded tapes. American record manufacturers seem reluctant to enter into yet another recording format. Besides, they haven't expressed a particularly favorable attitude towards DAT from the start.

Prerecorded DATs are likely to come primarily from small, independent companies and imports until the format is strongly established in the U.S. marketplace. Because of the weaker distribution networks for these sources, it is entirely possible DAT may die a premature (and unnecessary) marketing death because of a lack of software (prerecorded tapes). That would certainly be a shame.

Figure 7-1 *This is a typical automotive CD player.*

Of course, automotive DAT playback-only decks can also play tapes made on home DAT decks, just as many consumers make their own analog audio cassettes to play on their car stereo systems. In the case of DAT, however, home decks are still rather scarce and very expensive.

No one can really predict the future of DAT, although I'll present a few common theories in Chapter 9. In the meantime, this section will look at some of the automotive DAT decks now (or soon to be) on dealers' shelves. For purposes of comparison, an automotive CD player is shown in Fig. 7-1.

Condensation

Because of the temperature extremes encountered in the automotive environment, condensation can be a problem. If dew forms on the tape guides, it can cause tape jamming and other possible damage to both the tape and the player's mechanism if playback is attempted before the condensation has been cleared. For this reason, all automotive DAT decks include moisture sensors. If an unacceptable amount of condensation is detected, the cassette cannot be played until the tape guides have dried off. In many models, the cassette will be ejected immediately if it is loaded while the condensation indicator or "dew light" is on. Typically, it takes about 30 minutes for condensation to clear enough so that the tape may be safely played.

You will most likely encounter condensation problems in the winter months. The temperature changes that occur when the heater is turned on and off can cause considerable amounts of condensation to form, especially on metallic parts. To a somewhat lesser extent, this can also happen in the summer when the air conditioner is used. Sudden climatic changes, especially extreme temperature changes, are the culprit.

Kenwood KDT-99R

The KDT-99R FM/AM DAT player from Kenwood is shown in Fig. 7-2. Besides playing digital tapes, this unit also includes a complete stereo FM/AM radio receiver.

A particularly nice feature of the KDT-99R is the nine-stage tape remaining/loading indicator. This is a strip of nine joined LED indicators. During the playing of a tape, the number of lit LEDs indicates the approximate time remaining on the tape being played. At the beginning of the tape, all nine indicators are lit. As the tape progresses through its length, the LEDs are turned off, one by one, starting from the rightmost indicator. If only the left two indicators are lit up, you know that less than two-ninths of the tape remains to be played.

This system is inevitably less accurate than a true numerical readout, but the data can be read more easily at a quick glance from a bar graph display like this. A standard numerical timing readout display is also provided on this machine if more precise timing information is required.

What happens if you insert a DAT cassette that hasn't been fully rewound? No problem. The deck's circuitry will not be fooled into thinking the half-wound tape is starting at the beginning. The tape remaining indicator takes advantage of the time code data included in the subcode field of every DAT recording. In effect, the tape itself "tells" the player how much time remains on the tape.

When a tape is being loaded into the player, all nine of the

Figure 7-2 Kenwood KDT-99R automotive FM/Am DAT player.

tape remaining indicator LEDs blink on and off. This blinking continues until the player retrieves the necessary data from the tape.

Personally, I think this is a very handy feature. It certainly isn't essential, but it is very useful. I think this particular feature would be even more handy on a home recording deck. The user would not have to worry about running out of tape unexpectedly while recording important material.

Ordinarily, when the DAT cassette plays through to its end, the KDT-99R automatically rewinds the tape to the beginning and ejects the cassette. Sometimes, however, repeat play is desirable; especially while driving, when you can't conveniently or safely change tapes. For this reason, the KDT-99R also includes a "repeat all" button. When this switch is activated, the automatic cassette eject mechanism is turned off. When the tape plays through to the end, it is automatically wound back to the beginning and starts to play again.

There is also a "repeat one" switch on the KDT-99R. When this button is pressed, the current program (or selection) will be played repeatedly. When the song ends, the KDT-99R will automatically rewind and relocate the beginning of the song and play it again. Once again, subcode data is used to locate the appropriate points on the tape.

The KDT-99R can also be programmed to search for and start playing a specific program located anywhere on the tape. Up to 99 individual programs can be accessed by the KDT-99R. Program numbers of 100 or higher cannot be searched by this machine.

The search process involves the player automatically fast forwarding or rewinding to the appropriate point on the tape. Access time is not instantaneous. The tape must be physically wound to the appropriate spot and the subcode data must be read. The search operations function quite rapidly—about 200 times as fast as the ordinary playback speed. Of course, if you start the tape at the beginning and search for a program near the end of the tape, it will take longer for the search function to work than if you searched for a closer program.

On some tapes, the program numbers may not be recorded in consecutive order. This is likely to be the case, for example, when a user has made his own compilation tape from several prerecorded tapes without redefining the program numbers. If the program numbers are out of order, it is reasonable to expect the searching operation to take somewhat longer than normal. With consecutive program numbers, the player's search mechanism can reasonably "guess" where to find the target program. When the program numbers are nonconsecutive, the program being searched for could be located almost anywhere along the entire length of the tape.

If you tell the KDT-99R to search for a nonexistent program number, it will search until it is clear that the target program cannot be found. The tape will then be rewound to the beginning and normal (straight-through) playback will start. On some tapes, there may be no program numbers included in the subcode fields. Obviously, the KDT-99R's search functions can't be expected to work with such a tape.

The KDT-99R will also search for the next program, without the user having to know the specific program number. Kenwood calls this function "music search." This is basically the "next track" function from the basic DAT standards. It is similar to the "next track" function found on virtually all current CD players.

An index search function permits the user to preview the selections recorded on the cassette. The first 15 seconds or so of each program or song is played, and then the machine automatically fast forwards to the beginning of the next selection. The index search operation can be discontinued at any time simply by pressing the index search button a second time, or by pressing the play button.

A five-digit time display is included on the front panel of the KDT-99R. This display can be set for either total tape time, or program time. In the total time mode, the readout will display how much time has elapsed from the beginning of the tape. This data is read from the tape itself, so the display will be correct even if an incompletely rewound cassette is loaded into the player. Of course, this numerical display can provide a more precise and accurate readout than the nine LED bar graph indicator discussed earlier in this section. Obviously, the program time display indicates how much time has elapsed since the beginning of the current selection.

It is possible that some DATs may not include all of the necessary timing data in their subcode fields. In such cases, the KDT-99R's time display will function as a tape running indicator, similar in concept to the mechanical "footage" counters found on most analog cassette recorders. Under these circumstances, of course, the unit will have no way of knowing if an incompletely rewound cassette has been loaded into the player.

Correct operations in any DAT player depend heavily on the actual contents of the recorded subcode data fields on the tape being played. If a portion of the cassette happens to be completely blank (with no formatting data at all), the KDT-99R will automatically fast forward the tape to a point where a properly recorded signal can be read.

The KDT-99R can play back tapes using either the 44.1- or 48-kHz sampling rates. This unit does not support the optional 32-kHz sampling rate used in slow-speed recordings.

The manufacturer's specifications for the KDT-99R are very impressive, especially when compared with the specifications for most standard analog cassette decks. The wow and flutter is below any measurable limit. This is basically an analog specification that doesn't have any real meaning for digital recorders. This analog specification seems to be given for digital equipment simply because it looks good and most consumers expect to see it. If no wow and flutter specifications are given, some potential customers who don't understand digital recording might assume that the manufacturer is trying to conceal inferior performance in this area.

The frequency response of the KDT-99R is specified at 10 Hz to 20 kHz (20,000 Hz) ± 1 dB. The total harmonic distortion (THD) for this unit is rated at 0.005% at 1 kHz.

The KDT-99R automotive DAT player has a dynamic range rating of 92 dB. The signal:noise ratio for this player is also rated at 92 dB.

You should be aware that all of the specifications given here are for the DAT player only, not for the built-in radio receiver. A separate set of specifications are provided for this (strictly analog) section of the KDT-99R. Of course, these analog receiver specifications are irrelevant in this book, so we won't go into them here.

Clarion Audia 8100 AM/FM DAT player

The Clarion Corporation of America offers the Audia 8100 which is an automotive DAT player with a built-in AM/FM stereo radio, with CD changer control capability. Clarion claims this last feature is unique (so far anyway) to their product.

Multifunction buttons are used to operate the radio, DAT

player, and CD changer. By assigning multiple functions to most control buttons, Clarion has been able to give the Audia 8100's front panel a very clean, uncluttered appearance.

Supported DAT functions for this unit include tape selection repeat, intro tape scan, or return to the beginning of the tape. The Audia 8100 also has automatic program control (APC) capabilities, allowing the user to locate any selection on the tape with the touch of a button.

Here we see an example of different manufacturers using different terminology for the same basic functions. The same functions are employed in the Kenwood KDT-99R discussed above, but different names are used as shown in Table 7-1.

The Audia 8100 supports all three of the standard sampling rates defined in the basic DAT specifications—48, 44.1, and 32 kHz. In the CD mode, the Audia 8100 can operate an external CD changer, such as Clarion's Audia 6100. The control unit provides a number of convenient features including

- Disc repeat,
- Track repeat,
- Intro scan, and
- Automatic program control.

Notice that these special CD control functions mirror similar functions for the DAT player section.

The Audia 8100 is a fairly expensive piece of equipment, especially for an automotive installation. The list price at the time of this writing was almost $2300.

Needless to say, theft could be a problem with this unit. Fortunately, Clarion includes some built-in security measures in the Audia 8100. They call their security system "CATS," or "computer-controlled antitheft system." Basically, when the owner first installs the Audia 8100, he programs in his own individual

Table 7-1 Terminology used for basic functions

KDT-99R	Audia 8100
Repeat one	Tape selection repeat
Index search	Intro tape scan
Repeat all	Return to beginning
Program search	Automatic program control

Automotive DAT Decks 187

Figure 7-3 Clarion Audia 8100 automotive AM/FM DAT player.

Table 7–2 Specifications for the Audia 8100.

Wow and flutter	Below measurable limits
Frequency response	20 Hz-20 kHz ± 1.0 dB
THD	0.05% at 1 kHz
Dynamic range	92 dB
Signal:noise ratio	92 dB

customized security code. If the unit is removed from the vehicle for any reason, it cannot be played until the same security code is entered into it. This should be quite effective in reducing the resale value of stolen units.

The Audia 8100 is shown in Fig. 7-3, along with a listing of some of the unit's features. For the DAT player section, the excellent specifications for the Audia 8100 are similar to those for the KDT-99R (see Table 7-2).

Mitsubishi DT10 DAT player

Unlike the KDT-99R or the Audia 8100, the DT10 from Mitsubishi (Fig. 7-4) is a straight DAT player. This unit does not include a built-in AM/FM tuner. It does, however, include an auxiliary radio/tape input switch, permitting use with RCA-type pre-amp output radios, analog cassette players, or CD players.

The DT10 uses LCD instead of LED readouts. LCDs tend to be easier to read in bright sunlight. LCD readouts also consume significantly less power than LED displays of the same size. On the other hand, LCDs are generally more difficult to read in low-level light. All three of the standard sampling rates defined in

188 DAT Equipment

Figure 7-4 Mitsubishi DT10 automotive DAT player.

Table 7–3 Specifications for the DT10.

Wow and flutter	N/A
Frequency response	20 Hz-20 kHz
THD	0.01%
Dynamic range	85 dB
Signal:noise ratio	90 dB

the basic DAT specifications are supported by the DT10 DAT player—48, 44.1, and 32 kHz.

Once again, we have different names used for certain functions by different manufacturers. Mitsubishi refers to programmable repeat modes, sequential program scan, and "music program sensor" to jump ahead to the beginning of the next song (or the previous song) on the tape.

The manufacturer's specifications for the DT10 automotive DAT player are shown in Table 7-3.

HOME DAT RECORDERS

As of mid-1990, there were still very few models of DAT recorders on U.S. dealers's shelves (except for the grey market dealers discussed earlier in the preceding chapter). A number of manufacturers have shown prototypes, and promise to have some models out soon.

Addresses of companies who are currently manufacturing or have indicated some intent to manufacture DAT equipment will be listed in the Appendix of this book. In many cases, these companies are currently manufacturing DAT equipment for foreign markets only. A few of these companies may not have any definite plans to market DAT equipment in the U.S. in the near future. Others are already actively tooling their factories to start production on DAT products for the U.S. market. A few of these manufacturers have already released DAT products in the U.S. Some of these products are discussed in this chapter.

To the utter frustration of many eager audiophiles, many promises of legitimately sold U.S. DAT recorders have been knocking around for several years. The big holdup, of course, has been the legal tangle described in some detail in Chapter 6. Fortunately, it seems the SCMS compromise, described in the last chapter, has probably resolved most of the major legal conflicts. This should allow more and more manufacturers to start making good on their promises of marketing DAT recorders to U.S. consumers.

Casio DA-2

Possibly the first DAT recorder to officially appear on the U.S. market was the Casio DA-2, although even here Casio is hedging its bets somewhat. They are marketing the DA-2 to the professional market, not to general consumers. The price tag (just under $1500, list) would place this machine out of the reach of all but the most devoted audiophiles anyway.

The DA-2 is designed to be a portable unit, weighing just 3 pounds, including the supplied onboard battery. The battery's charge is good for about 2 hours of use, or one standard DAT cassette at the normal speed.

Thanks to the small size of the DAT cassette itself, the DA-2 is a very small unit. It measures a mere 10 inches by 6 inches by 1.75 inches. This is a top-loading machine. The cassette is placed in a well on the top of the recorder. This cassette well has a damped, motor-driven elevator door.

The DA-2 does not support the optional 32-kHz sampling mode. Only the 48- and 44.1-kHz sampling rates are used by the DA-2. This means you cannot use the optional slow speed with this particular unit.

There is no user control to set the sampling rate. Such a control is completely unnecessary. The choice of sampling rate is always made automatically by the DA-2. When recording, the

48-kHz sampling rate is always used, since the DAT standards prohibit consumer recording at the 44.1-kHz sampling rate. When playing back a tape, the DA-2 automatically determines whether it was recorded with a 44.1- or a 48-kHz sampling rate, and automatically adjusts itself accordingly.

System information is supplied through a small LCD readout panel on the front of the recorder. No back lighting is used for this display window, and it may be somewhat difficult to read under certain lighting conditions, especially considering its rather small size. A larger display window would have required a larger case for the recorder and greater current consumption, thus reducing its portability, so this was a trade-off design. In practice, however, it probably wasn't too drastic a trade-off. How often will the average user try to record without sufficient lighting? The DA-2's display window should be perfectly adequate for most practical applications.

The same readout is used for various functions, which can be selected by simple front-panel switches. The same LCD unit displays both the time and index information and level metering for recording.

Four different modes are employed to display time information on the DA-2. They are

- Real absolute time (from beginning of tape),
- Elapsed program (or track) time,
- Remaining tape time, and
- Numeric counter.

The numeric counter mode is functionally similar to the common mechanical and electromechanical counters found on most analog tape recorders. The counter can be reset to zero at any time during the recording or playing of a tape. The time display mode on the DA-2 is switch selectable, except when no appropriate timing data subcodes have been recorded on the tape being played.

Level metering, a multisegment display, is used for recording (and playback). The display ranges from -40 dB to 0 dB. If a level greater than 0 dB is presented to the recorder, the word "over" is displayed. With a digital recorder, it doesn't make any difference how much the signal is over the 0-dB limit. Once the signal level goes over 0 dB, very severe distortion occurs. Below

0 dB there is very little measurable distortion (it is well below audible levels). The 0-dB point is a hard switching line for digital recording. This is in contrast to analog recording where the distortion level steadily increases with increases in the signal level, even below 0 dB. On an analog recorder, 0 dB is a fairly arbitrary point of compromise.

With analog tape recorders, the general practice is to set the input level controls so that the maximum peak levels indicated by the VU meters come as close to 0 dB as possible. Signal levels above 0 dB in an analog recorder increase in distortion with further increases in level. Small, short-duration overpeaks generally don't have a particularly noticeable audible effect; if the signal occasionally peaks above the 0-dB level, there is no real problem.

Digital recording requires somewhat different recording practices. Any signal peaking above 0 dB is flatly clipped at 0 dB, resulting in distortion, which can often be quite severe. This effect is illustrated in Fig. 7-5. The squaring off of the peaks of the wave form creates a number of strong false harmonics, or distortion. Just a little bit of this would be highly objectionable. Therefore, with a digital tape recorder, it is always best to set the record signal level controls to leave some room below 0 dB. A good rule of thumb is to set the recording level so that the music peaks never give a reading higher than about -6 dB.

Several possible subcode modes are available for recording tapes on the DA-2. Generally, the mode switch is set to "normal," where standardized subcode date will be recorded onto the tape. There are also three additional modes used for adding and editing various ID codes for special purposes, such as providing index numbers like those used on some CDs.

Certain programs or selections on a tape can be programmed to be automatically skipped on playback. Also, various programs (selections or songs) can be rapidly accessed during playback. The DA-2 can be instructed to look for a specific program number. It will then fast forward or rewind the tape to find and play the appropriate selection. The tape is moved at a much higher than normal speed in the search mode (about 150 to 200 times the normal playing speed).

Most DAT recorders automatically rewind the cassette to the beginning when it is played to the end. The DA-2 normally does this too, but not if there is more than 9 seconds of silence at the end of the tape. In this case, at the end of the cassette, the DA-2

192 DAT Equipment

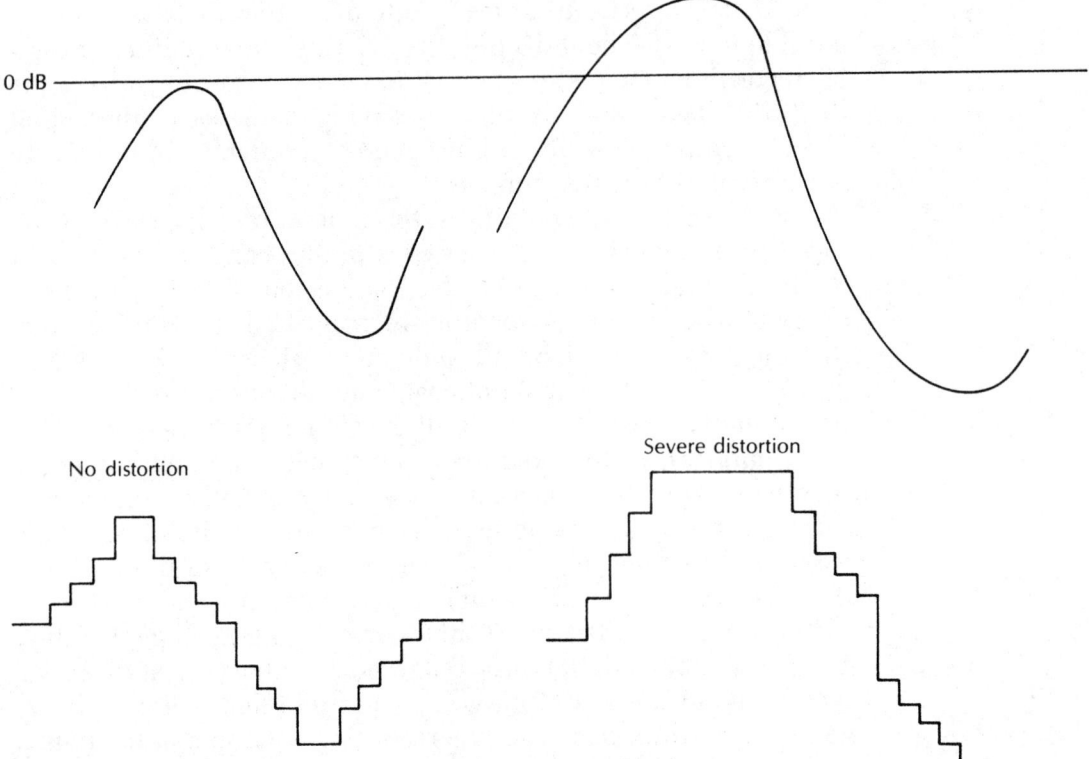

Figure 7-5 Digital recording with a signal level above O dB results in severe clipping distortion.

will rewind the tape only to the beginning of the blank portion of tape and stop there. This is very handy for making recordings discontinuously. You can play back what you've recorded so far, then have the recorder automatically locate the exact point where you left off recording, so you can conveniently pick up right where you left off.

As with all DAT equipment, the specifications for the Casio DA-2 are quite impressive (see Table 7-4).

Table 7–4 Specifications for the DA-2.

Frequency response	10 Hz to 20 kHz ±1 dB
THD	0.01% at 1 kHz
Dynamic range	80 dB
Signal:noise ratio	85 dB

Kenwood I-Z9

Several manufacturers have announced DAT recorder prototypes that were not yet being marketed at the time of this writing. In most cases, it can be expected that the prototype will be very similar to the commercially released unit, except possibly for some minor cosmetic differences.

The L-Z9 is a prototype DAT recorder from Kenwood. Unlike the DA-2, this unit is designed to use all three of the sampling rates defined in the basic DAT standards:

- 48 kHz (record and playback),
- 44.1 kHz (playback only), and
- 32 kHz (record and playback).

The L-Z9 is a deck-type recorder, rather than a portable machine like the DA-2, and it is somewhat larger in size. It is designed to be added as part of an existing home sound system. A separate amplifier and speakers are required for playback.

The L-Z9's physical dimensions are approximately 17 inches by 12.75 inches by 4 inches. It is intended to be used in a fixed location and is ac powered. Battery power is not supported with this unit. The unit weighs about 7.5 kg, or a little under 17 pounds.

The stated specifications for the L-Z9 are shown in Table 7-5.

The L-Z9 DAT recorder is designed for use with a remote control for added convenience. The operations and functions of the L-Z9 are similar to Casio's DA-2 DAT recorder.

Nakamichi 1000

The Nakamichi 1000 DAT system, shown in Fig. 7-6, is unquestionably designed to be a real "top-of-the-line" DAT recorder.

Table 7-5 Specifications for the L-Z9.

Wow and flutter	Below measurable limits (less than ±0.001%)
Frequency response	2 Hz to 20 kHz ±0.05dB
THD	0.002% at 1 kHz
Dynamic range	90 dB
Signal:noise ratio	90 dB

194   DAT Equipment

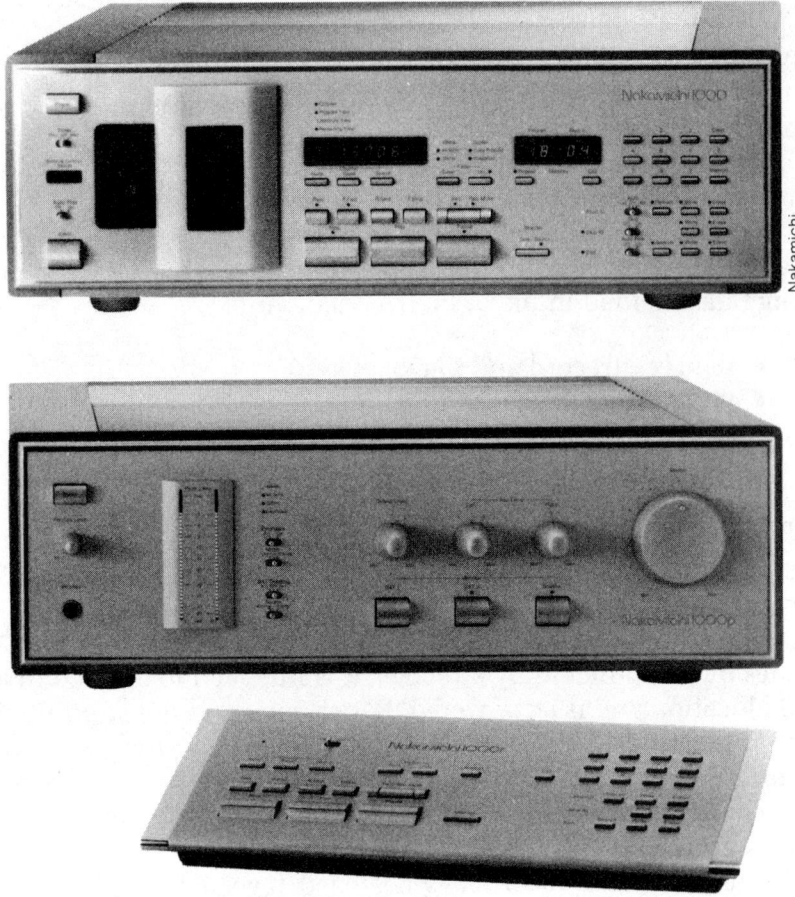

Figure 7-6   *Nakamichi 1000 digital audio recording system.*

The basic recording system is divided into two separate units. The Nakamichi 1000 digital audio recorder, containing the basic transport/recorder section, is shown in Fig. 7-7, and the Nakamichi 1000p digital audio processor is illustrated in Fig. 7-8. This section contains the circuitry to perform the actually digital signal processing (analog-to-digital and digital-to-analog conversion).

Nakamichi split up their recording system into two sections like this to maximize performance and facilitate future system upgrades and expansions. Unlike most DAT recorders, the Nakamichi 1000 digital audio recording system is specifically designed to be fully expandable through the use of modular, plug-in circuit boards. This minimizes the possibility of the system

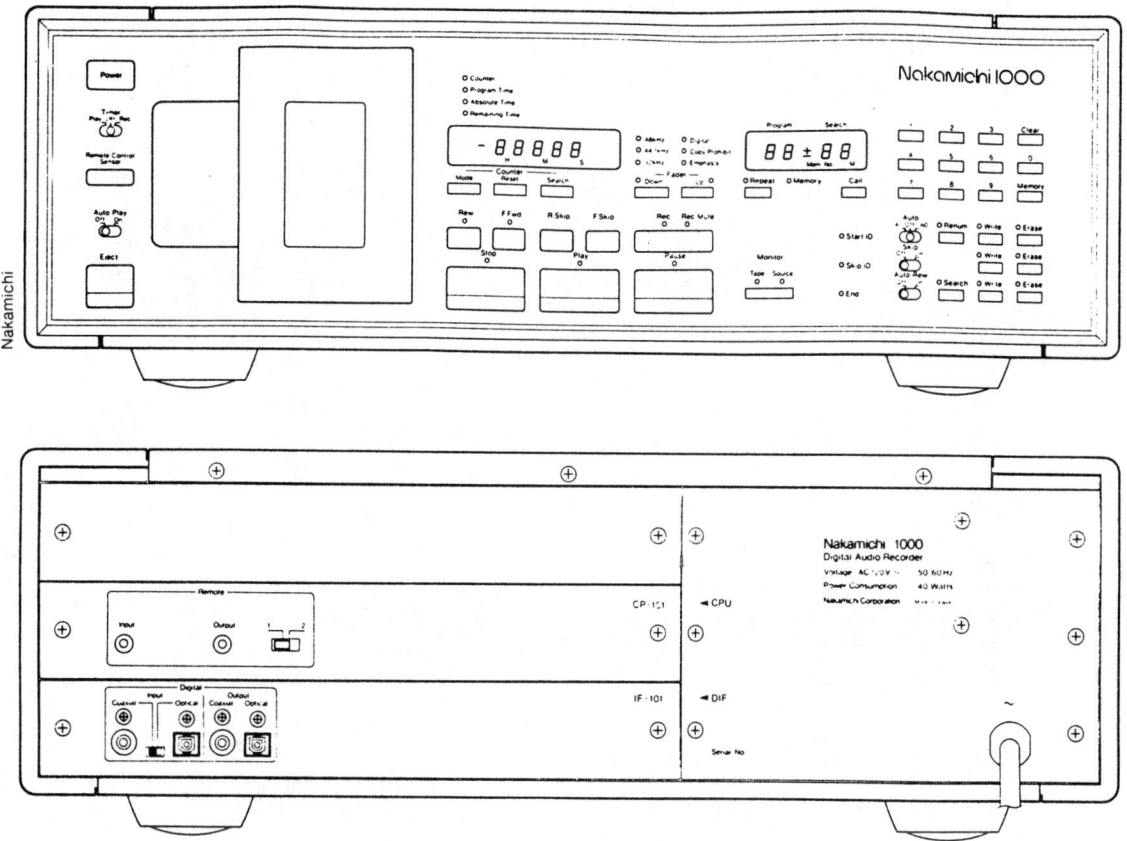

Figure 7-7 Nakamichi 1000 recording unit.

being made obsolete by new advances in technology. You can upgrade the system by plugging in a new expansion card.

The basic Nakamichi also includes a remote control, the Nakamichi 1000r. This unit is illustrated in Fig. 7-9. The remote control for the Nakamichi 1000 digital audio recording system is available in both wired and wireless versions.

The Nakamichi 1000 recorder and the Nakamichi 1000p digital processor may be purchased separately. The digital processor unit can be used to control either one or two recorder units, allowing users to expand their system by buying a second 1000 recorder unit. They would not need to pay for a redundant digital processor.

The Nakamichi 1000p digital processor unit can also be purchased as a stand-alone (without the 1000 recorder) converter for

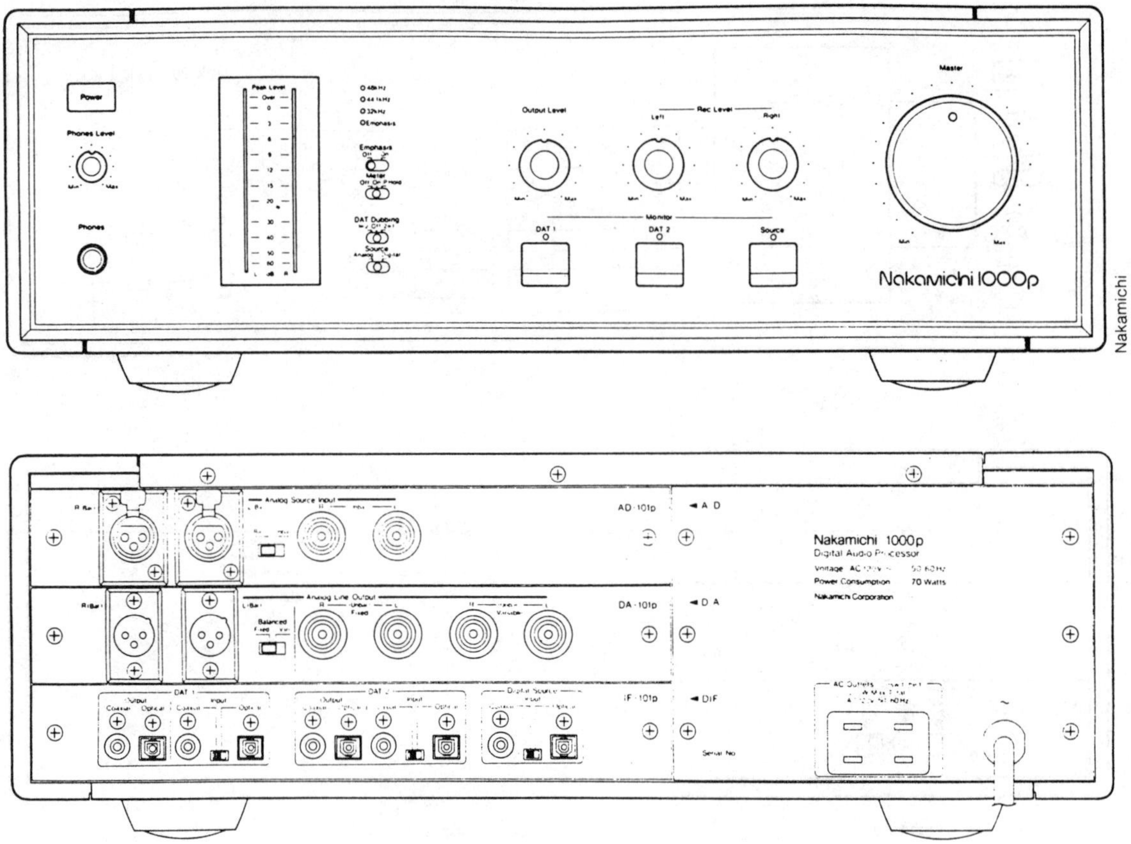

Figure 7-8 Nakamichi 1000p digital processor unit.

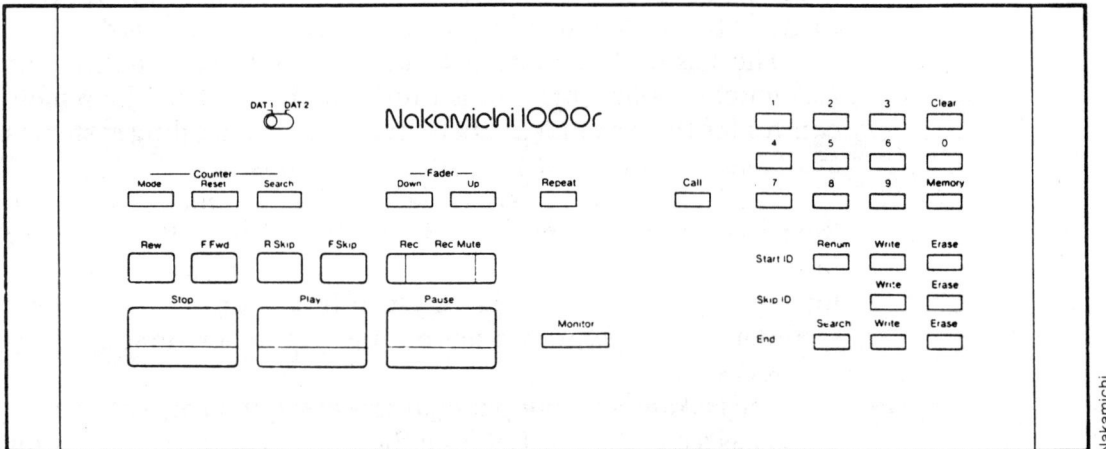

Figure 7-9 Nakamichi 1000r remote controller.

use with other DAT recorders or CD players that provide digital interfaces. Both coaxial and optical connections are supported on this machine.

A "professional" version of the Nakamichi 1000 digital audio recording system is also available, although it is hard to imagine many amateurs buying any version of this system for casual home use. The basic system (including the 1000 recorder, the 1000p digital processor, and the 1000r remote control) lists for $11,000. The 1000p digital processor unit by itself sells for $5,100. The 1000 recorder unit (with a 1000r remote control) has a list price of $5,900.

The main differences of the "professional" model are standard 19-inch EIA rack-mount handles, and inputs and outputs that conform to studio standard line levels. The "professional" 1000 recorder is illustrated in Fig. 7-10, and the "professional" 1000p digital processor is shown in Fig. 7-11.

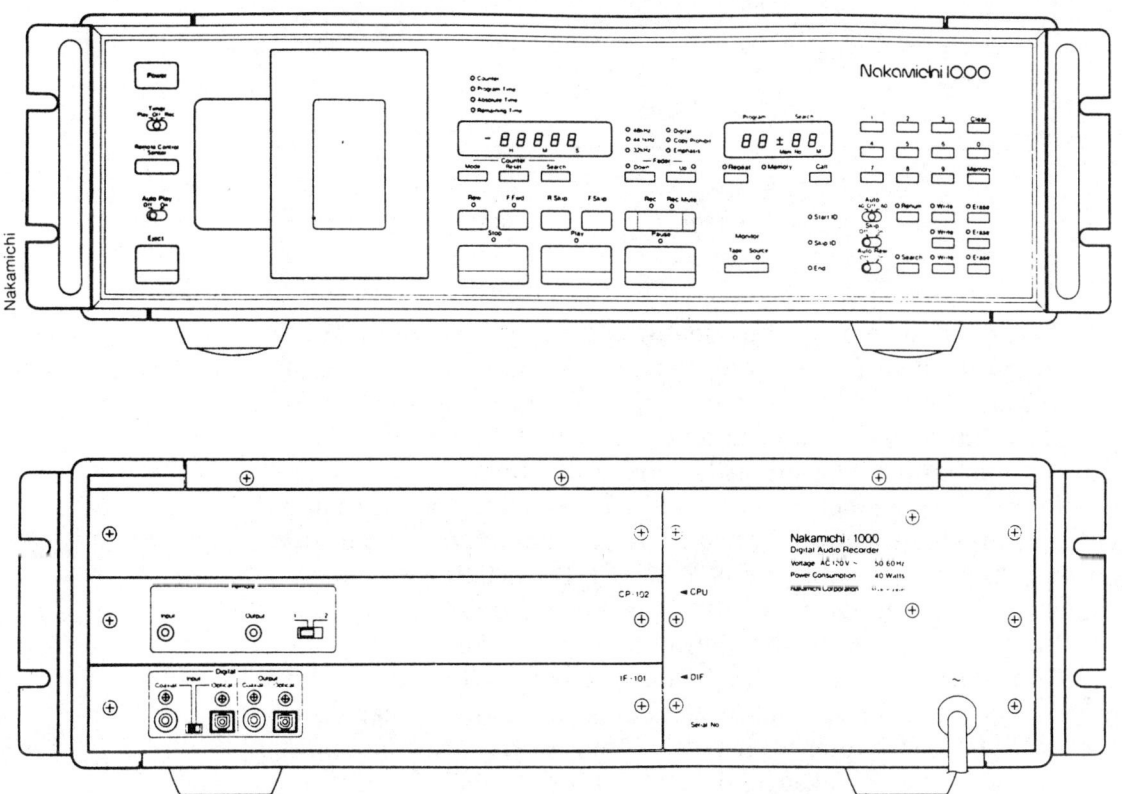

Figure 7-10 Nakamichi 1000 "professional" recording unit.

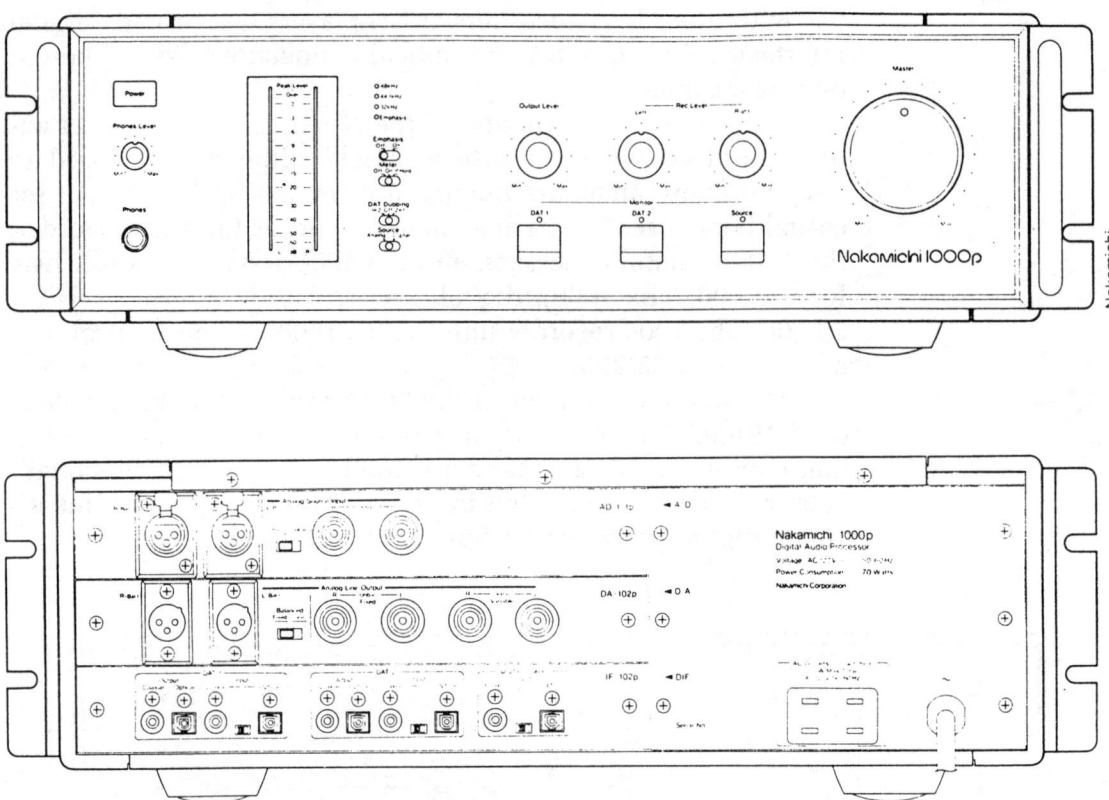

Figure 7-11   Nakamichi 1000p "professional" digital processor unit.

There are several unique features in the Nakamichi 1000 digital audio recording system. Perhaps the most surprising of this system's many features is the capability of recording digital signals at the 44.1-kHz sampling rate, despite the normal restrictions of the basic DAT standards. Niro Nakamichi explained his position in a press release, saying, "We've maintained a dialog with the music industry because we're not looking for a confrontation. There will be objections, I expect. But whatever the consequences, intentionally crippling the capabilities of a machine like this was a compromise I simply couldn't make."

Considering the hard-line attitude taken by the RIAA and most major record manufacturers against the copying capabilities of DAT (see Chapter 6), it will be interesting to see just how they respond to the Nakamichi 1000's ability to make copies directly at the 44.1-kHz sampling rate. Of course, the Nakamichi 1000 also supports full record and playback capabilities at the standard 48- and 32-kHz sampling rates.

On the one hand, this system is priced well beyond the reach of most nonprofessionals. On the other hand, if the RIAA doesn't fight this machine, they'll have a harder time legally opposing a later, less expensive system with similar capabilities.

As I said, it's definitely going to be interesting to follow the developments here. The entire copying controversy is discussed at some length in Chapter 6.

Another fairly unique feature of the Nakamichi 1000 recorder is an improved tape transport system called "FAST" (fast access stationary tape guide transport), which uses stationary tape guides for improved tape travel precision. Better precision in the tape path results in a lower digital data error rate and improved long-term reliability of both the tape and the mechanical parts of the recorder.

Tape loading is also faster with this system. Nakamichi claims tape loading can be completed in less than 2 seconds.

For nonsearch fast forwarding and rewinding, the tape is put into a special "half-load" position, in which the tape is moved out of contact with the rotary head drum. Fast forwarding and rewinding using the half-load position can be 400 times as fast as the normal playing speed, or twice the fast winding speeds of standard DAT mechanisms. Once again, moving the tape away from the head drum reduces wear and tear on both the heads and the tape itself.

Most DAT recorders have two record/play heads mounted in the rotary head drum. The Nakamichi 1000 doubles this to four heads, permitting true off-the-tape monitoring of the recorded signal as it is recorded.

The Nakamichi 1000 digital audio recording system uses an 8 times oversampling, 20 bit digital-to-analog converter along with customized calibration ROM (read-only memory) chips. The ROMs hold customized data to compensate for any inaccuracies in the individual converter ICs used in that particular machine.

For professionals and serious semiprofessionals (and very well-heeled audiophiles), the Nakamichi 1000 digital audio recording system certainly has a lot to offer. Needless to say, the specifications for this device are excellent (see Table 7-6).

Table 7–6 Specifications for the Nakamichi 1000 D/A converter.

Frequency response	5 Hz to 20 kHz ±0.5 dB
THD	0.0015% at 1 kHz
Dynamic range	100 dB (minimum)
Signal:noise ratio	106 dB (minimum)

Table 7–7 Specifications for the Nakamichi 1000 A/D converter.

Frequency response	2 Hz to 22 kHz ± 0.5 dB
THD	0.003% at 1 kHz
Dynamic range	94 dB (minimum)
Signal:noise ratio	96 dB (minimum)

The measurements for the D/A converter section are all made at the 44.1-kHz sampling rate. It would be reasonable to expect similar (almost identical) measurements at the 48-kHz rate. Table 7-7 shows the specifications for the Nakamichi 1000 A/D converter.

The measurements for the A/D converter section are all made at the 48-kHz sampling rate. It's a bit curious the Nakamichi has presented the specifications for their recording system in this fashion. Ordinarily, the A/D converter specifications would not be appropriate for the 44.1-kHz sampling rate. Since this is (according to the usual DAT standards) a playback-only speed, the A/D converter section, which is used for preparing signals for recording, would not usually be employed at the 44.1-kHz sampling rate. However, the Nakamichi 1000 system does permit recording at 44.1 kHz.

I really don't know why, under the circumstances, they chose to use different sampling rates to determine the specifications for the D/A converter and A/D converter sections.

The Nakamichi 1000 recorder measures 17.125 inches by 5.22 inches by 14.56 inches, and weighs 35 pounds, 4 ounces. The Nakamichi 1000p digital processor also measures 17.125 inches by 5.22 inches by 14.56 inches, and weighs 38 pounds, 9 ounces.

DIGITAL AUDIO AND SPECIFICATIONS

While I have reported the key manufacturer's specifications for the equipment described in this chapter, I'd like to take a moment to add a word of caution against reliance on published specifications. They don't always mean as much as you might think.

A digital recording medium generally offers excellent specifications when compared with similar analog recording equipment. However, this may be somewhat misleading in many practical cases. Specifications that are critical in defining differences between analog equipment may be irrelevant for digital devices; while other factors, unique to the digital format, may not be measured in standard specifications.

An obvious example of inappropriate specifications is the "wow and flutter" specification. (These terms are defined in Chapter 2.) Wow and flutter are of great importance for analog turntables and tape recorders, but such minor fluctuations in speed are utterly irrelevant in a digital recording. Wow and flutter will always be below measurable limits (or the minimum level indicated by the test equipment used) for any DAT recorder or other digital device. There is no real reason for manufacturers to include this specification for digital equipment, but it is almost always included. Presumably, this is because such specifications look terribly impressive and make effective advertising.

A more crucial area in digital specifications is the distortion rating. This is normally given as total harmonic distortion (THD). Once again, all DAT recorders have excellent values in this area. Ratings of 0.05%, or even less, are typical for digital audio equipment.

But does this mean the equipment is truly distortion free? Not necessarily. The THD measurements for digital audio components are made in the same way as for analog audio components. Digital distortion is, however, quite different from analog distortion.

For example, standard THD ratings are weighted to account for the fact that analog equipment shows increasing levels of distortion with increasing signal levels. In a digital system, on the other hand, the distortion level increases as the signal level decreases. Thus, the digital distortion may be underreported because of the way the measurements are weighted.

In fact, digital distortion may be more acoustically objectionable than analog distortion. In an analog sound system, the distortion increases during louder passages. The higher volume of the music itself can help mask some of the distortion effects.

They are still there, but the human ear doesn't really notice them as much.

In a digital audio system, the highest distortion levels appear during the quietest passages of the music, so the distortion effects will not be masked; they will tend to be quite apparent.

All in all, many currently used audio specifications don't appear to be particularly appropriate to DAT recorders, or other digital sound equipment. New measurement techniques and specification standards need to be devised to meet the changing needs of the digital age.

The digital recording process is fairly consistent in terms of reproduced sound quality. The audible differences between DAT recorders are likely to be fairly subtle. To most people's ears there will be little or no difference in the sound between different models of DAT equipment.

But by all means, try to listen to any equipment before you buy it. If you detect a marked difference between the sound of two competing units, it doesn't matter what anyone else says. Purchase the one that sounds better to your own ears. When it comes to spending money on your sound system, your opinion is all that ultimately matters. You're the one who's going to be listening to the system, aren't you?

For most of us, when shopping for a DAT recorder or player, there isn't much point in getting hung up on the specifications, or any slight differences in sound quality. Instead, you will want to confine your comparisons to three areas:

- Price,
- Durability and quality of construction, and
- Features.

The price issue is pretty obvious. For the time being it is also pretty irrelevant for DAT equipment. It's all either expensive or very expensive. Price should not be the only consideration, however. The question is value, not literal cost. This brings us to the question of durability and quality of construction. How well is the unit put together? The early DAT recorders and players are all true top-of-the-line units, and all seem to be well and solidly assembled.

In shopping for any electronic equipment, including DAT recorders, the case is generally a pretty good indicator of how well a piece of equipment is constructed. Does the case look and feel solid, or does it look like it will crack or shatter if it is dropped? (Certainly you want to avoid dropping your DAT recorder, but, after all, accidents do happen.) If possible, all other factors being equal, it is usually a good idea to opt for a machine with a die-cast metal housing. This is better and sturdier than thin sheet metal or plastic cases. However, such a case will add to the weight and cost of the unit.

Finally, there is the area of features. Think about what each special feature does and what it means to you, if anything. Some people are unduly impressed by a lot of snazzy buttons, whether they know what they're for or not. These people often waste their money paying for features they don't want or need, and won't use.

You can bet that different DAT recorders and players will soon offer a mind-boggling variety of programmability functions and other special gimmicks and tricks. Certain features you might use every day, while someone else will never have a use for them. Consider your own individual needs and temperment. Does the feature sound worthwhile to you? If you're not going to use a particular feature, why pay extra for a machine that supports that feature?

Overall, unless you buy a very cheap model, that is obviously flimsy and poorly designed, you're not going to get stuck with a piece of junk when you buy a DAT recorder or player. It's just a question of being slightly satisfied or totally satisfied. Consider your own individual needs and preferences. They count for far, far more than any amount of advertising hype.

8 ❖
Maintenance and Troubleshooting

THERE IS A CURIOUS PARADOX THAT ALWAYS COMES UP ABOUT electronic equipment and high technology. Very often, the more complex an electronic device is, the simpler it is. That is, while the circuitry and mechanical devices involved may be very sophisticated and advanced in design, high-tech equipment tends to be very easy to use, generally requiring very little maintenance.

High-tech devices tend to be surprisingly reliable. Repairs are rarely needed on most high-tech devices unless there is serious misuse of the equipment. Seemingly simpler devices are often considerably less durable in use.

For example, internally, a CD player is far more complex than a traditional turntable for analog LPs. The CD player has many more component parts and demands greater precision for its mechanical moving parts and its electronic components. Logically, it might seem that there would certainly be a lot more that could go wrong with the complicated CD player. In actual fact, however, most CD players are extremely reliable and tend to exhibit far fewer problems and breakdowns than typical LP turntables.

A DAT recorder is designed to be very easy to use. It's just a matter of inserting a cassette and pressing one or two buttons. The sophisticated digital circuitry inside the machine automatically takes care of all of the real work involved.

DAT recorders are also very, very reliable. Much of this reliability is largely built into the design of the DAT system. Occasional service calls, while always inevitable, are not likely to be particularly common with most DAT equipment.

For the best possible results with any type of high-tech equipment, it is always wise to follow a few simple general maintenance procedures. Taking the trouble to do a little simple maintenance can prevent many problems before they occur. A number of handy tips for keeping your DAT recorder running at its best will be given later in this chapter.

This chapter will also offer some suggestions for what you should do when things do go wrong (or seem to go wrong) with your DAT equipment. We will be concentrating here on user-level servicing only. Many common problems can often be easily corrected without calling a professional technician.

This chapter on DAT maintenance and troubleshooting will not deal with any actual electronic or mechanical repairs. Such advanced coverage is beyond the scope of this book. This volume is aimed at the general user of DAT equipment, not the professional electronic technician.

If you do not know what you are doing, do not open the case of any DAT recorder, or any other piece of sophisticated electronic equipment for that matter. You're liable to end up doing far more harm than good. All of the "repairs" discussed here are system repairs—all done without opening the equipment.

Troubleshooting, in general, is simply a matter of thinking about the problem and its symptoms logically and determining what could possibly be causing it. More often than not, with DAT equipment the problem is likely to result from incorrect settings of controls or incorrect system connections rather than internal mechanical or electronic problems within the DAT recorder itself.

The main aim of this chapter is to let the user know when he really needs to call in a service technician and when he can save himself some money and avoid wasting a trained technician's time with a trivial problem.

CARE OF DAT CASSETTES

Digital audio tape is enclosed in a plastic cassette, which protects it from dust, fingerprints, and other contamination risks. This protective housing simplifies the care of DAT, but it does not make the tape entirely invulnerable to all possible abuse. Nothing can do that. No matter how "idiot-proof" a system may be, you can always count on some idiot to find a way past all the safeguards to screw things up anyway. At least, the DAT cassette's housing protects against most legitimate accidents and misjudgements.

Like a video cassette, a DAT cassette has a hinged door that is automatically opened by a special mechanism within the machine to withdraw some of the tape for recording or playback. Never attempt to manually defeat the protective latch and open this door yourself. It's there for a reason. Leave it alone. Never touch the enclosed tape or attempt to pull it out of the cassette housing.

Don't ever try to open or take apart a DAT cassette. It is a precision device and you will almost certainly be unable to reassemble it properly. In the factory, DAT cassettes are assembled with specialized equipment. Even if it looks like you've put it back together correctly by hand, there may be subtle problems that could cause the tape to jam in your recorder and possibly do severe damage to some of the machine's delicate mechanical parts. There is also a very high risk of contamination of the tape's surface while the cassette is open.

If you can't restrain your curiosity and absolutely must take a DAT cassette apart to see how it looks inside, throw the tape and cassette away afterwards. Do not attempt to reassemble and reuse it. If you try, you're likely to end up with a repair bill that's a lot higher than the cost of a new DAT cassette.

Never attempt to physically edit a DAT tape using the manual cut-and-splice methods of analog tape recording. There are a number of reasons for this restriction. Obviously, you can't cut and splice a tape without touching it, which you should not ever do with DAT because of the risk of contamination (the oils on your fingertips can be quite harmful to the magnetic particles holding the data on the tape) and possible misloading.

Any splice results in a momentary change in the thickness of the tape (the thickness of the splicing tape is added to the regular thickness of the recording tape). This can affect the loading of the tape through the tape path, and again this could result in serious jamming problems. But perhaps the most important reason you should not attempt cut-and-splice recording with a DAT recorder is that it just won't work. In an analog tape re-

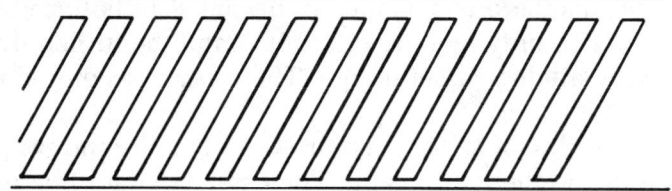

Figure 8-1 DAT signals are in the form of sequential diagonal tracks across the width of the tape.

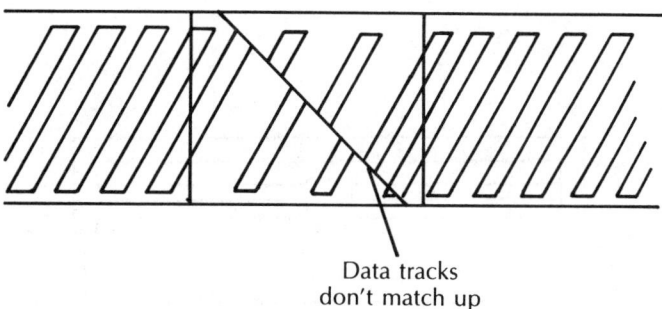

Data tracks
don't match up

Figure 8-2 DAT tapes cannot be edited using manual cut-and-splice methods.

corder, the signal is laid down along the length of the tape in a continuous line. The tape can be cut and rejoined at any point, and the playback head will be able to recover the edited signal.

In a DAT recorder, however, the data is placed on the tape in the form of sequential diagonal tracks across the width of the tape, as shown in Fig. 8-1. Figure 8-2 shows the typical results of cutting and splicing such a tape. Considerable data would be lost before the playback head could find and resynchronize itself to the recorded tracks.

All DAT editing must be done electronically; that is, two separate DAT recorders are required for the editing process. The tape to be edited is played back on the first machine, and the desired portions of the signal are rerecorded in the desired sequence onto the second machine. If the rerecording procedure is done entirely in the digital realm, there will be absolutely no degradation in the recorded signal quality. Even if the digital signal from the first machine is converted back into analog form and redigitized by the second recorder, the degradation in the sound quality will be minimal. It probably won't even be audibly noticeable, if care is taken in the rerecording process.

208 Maintenance and Troubleshooting

A DAT cassette is specifically designed to be inserted into the recorder in only one direction. It cannot be turned over like an ordinary analog cassette tape. There is no second side on a DAT cassette.

The top side of a DAT cassette has a small window, permitting you to see the tape inside. The cassette should be inserted into the recorder with this side up and the hinged door facing away from the user, as shown in Fig. 8-3. Never attempt to insert the cassette upside down, as illustrated in Fig. 8-4.

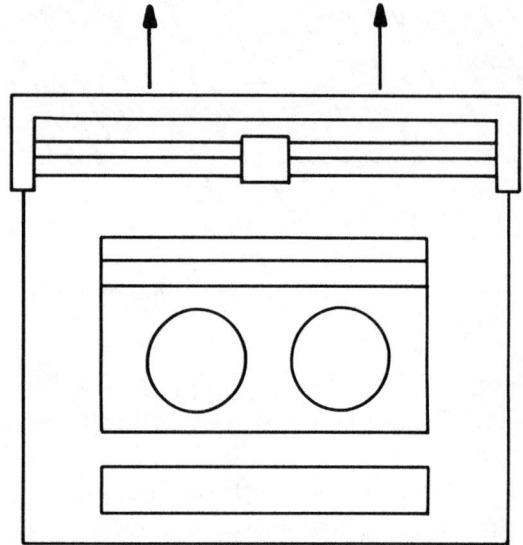

Figure 8-3 A DAT cassette must be inserted rightside up.

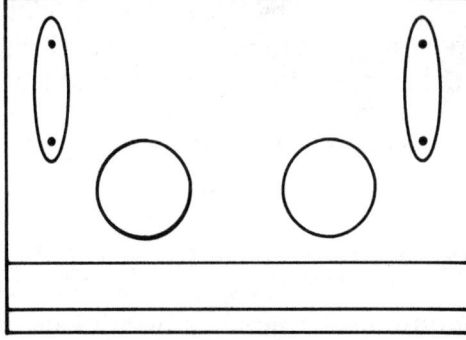

Figure 8-4 Do not attempt to insert a DAT cassette upside down.

When a DAT cassette is loaded into a recorder, part of the tape is automatically pulled out of the cassette and threaded through the tape path, including the partial wraparound the head drum. It is not advisable to repeatedly insert and remove a DAT cassette without playing it. The repeated loading and unloading could cause the tape to become slack or otherwise subject to damage.

If the label on a DAT cassette is partially peeling off, do not insert the cassette into your recorder. The label could come off completely and get caught somewhere in the machine's delicate mechanism. This could cause major jamming problems and possibly severe damage to your recorder. If you have a DAT cassette with a loose label, immediately remove the label altogether. If the label rips and portions of it remain firmly stuck to the cassette, this probably won't be a problem. If you can't get it off with your fingernails, it is unlikely to come off in the recorder's mechanisms on its own.

Undoubtably, suitable replacement labels for DAT cassettes will soon be made available. Similar replacement labels are already widely marketed for analog audio cassettes and videocassettes.

Condensation can be a serious problem with any type of cassette tapes. If dew forms on the tape, it can cause problems in the recorder's mechanism. You are most likely to experience condensation problems with a cassette when you bring it in from a cold environment to a warmer environment (such as the inside of a car to your home on a winter day). To avoid trouble, leave the cassette in the new environment for awhile allowing it to become acclimated to the ambient temperature and for any existing condensation to evaporate. Typically this will take 30 to 60 minutes.

Most DAT recorders will probably come equipped with internal moisture sensors and will automatically reject a cassette if there is condensation, but it is best not to take chances. If a DAT cassette feels cold to the touch, don't try to play it until it's had a chance to warm up.

While the hermetically sealed cassette used for DATs does a very good job of protecting the tape from dust and other environmental hazards, sensible storage can significantly reduce the chances of problems. Always store a DAT cassette in the case supplied with it. Don't expect the cassette housing to work miracles all by itself. Don't store DAT cassettes in dusty and dirty locations. Also avoid environments with excessive humidity or moisture. You certainly don't want to store your valuable DAT cassettes in a damp basement.

Avoid storing DAT cassettes at extremes of temperature, particularly excessive heat. Do not expose a DAT cassette to direct sunlight for prolonged periods. That is, don't leave it on a windowsill, or on the dashboard or the rear window parcel shelf of a car.

Try not to subject a DAT cassette to any strong shocks or vibrations. Avoid dropping it, especially from significant heights or onto a hard surface.

Never store any recorded tape near any source of magnetism. Remember, the signal on the tape is in the form of magnetic impulses. Exposure to a strong magnetic field could partially or totally erase your recording. Some common sources of potentially troublesome magnetic fields to watch out for include

- Speakers,
- Television sets,
- Transformers,
- Motors, and
- Telephones (especially with electromechanical ringers).

The telephone seems to be the one that catches most people off guard.

By applying a little common sense, a DAT cassette can last a long time and be good for more than 1000 playings.

CLEANING

Any tape recorder needs to be kept very clean to keep it in good operating condition. This is especially true for digital audio tape; even a small amount of contamination can completely wipe out large quantities of recorded data.

The mechanical design of a DAT recorder makes it rather awkward to clean using cleaning fluid and swabs. The heads are not as accessible as in an analog tape recorder. In many ways, we have the same situation as that found in most VCRs.

Special cleaning cassettes will certainly be made available. The cleaning cassette is loaded as if it were an ordinary DAT cassette. As it is played through, it cleans the heads and the tape path. Better cleaning cassettes will use some sort of cleaning fluid inserted through an access hole in the cleaning cassette's housing.

Be careful not to overuse a cleaning cassette, especially if it is of the dry (no cleaning fluid) type. Essentially, the cleaning cassette contains a pseudotape that functions something like a cloth wiped over the areas to be cleaned. The cleaning tape is designed so that dirt and dust particles will adhere to it. So what happens to the dirt once the cleaning cassette has been used? Nothing. It's still there on the "cloth." If too much dirt accumulates on the cleaning tape, it won't be able to pick up more and will start to smear dirt from previous cleanings all over the tape path and heads.

This point is often ignored by manufacturers of cleaning cassettes, and is rarely mentioned in their instruction sheets. This is true of any cleaning cassette, whether it is for VCRs, analog cassette recorders, or (presumably) DAT recorders.

Ideally, a fresh cleaning tape should be used each time, but this can get very expensive. As a rule, don't use a dry cleaning cassette more than about three times before discarding it. A cleaning cassette used with cleaning fluid should be good for about six cleanings, even though some manufacturers claim you can use it more than this.

If you can see any visible dirt on a cleaning tape, it is already far too dirty. Dispose of it immediately and use a new cleaning cassette.

How often should a DAT recorder be cleaned? There are no hard and fast answers to that question. It depends on how much the machine is used and for what purposes. In most professional studios, all recorders are routinely cleaned before each use. For home use, you probably don't have to be quite so strict. I haven't yet found any "official" recommendations on cleaning DAT recorders. But dirt is dirt, and magnetic tape is magnetic tape. It is reasonable to assume that the standards applied to analog tape recorders would apply just as well to DAT recorders.

Cleaning after every 10 to 20 hours is recommended for best sound quality and the maximum possible life span for your tapes and recorder. Most consumers clean their recorder's heads less frequently than this (which is, incidentally, one of the leading causes for problems requiring a service call). Table 8-1 presents a minimum cleaning schedule

Table 8-1 Cleaning schedule for DAT recorders.

Frequency of Use	Cleaning Schedule
> 5 hr/day	Twice weekly
2–5 hr/day	Once weekly
<2 hr/day	Monthly
l-10 hr/week	Every other month
1–10 hr/month	Every 6 months
<1 hr/month	Annually

Once a year is the absolute minimum recommended cleaning schedule. Even if the machine is left entirely unused, it will still get dirty from dust and other contamination particles in the air. If anyone regularly smokes near the recorder, double the cleaning schedule.

As far as any tape recorder is concerned, you probably can't clean it too frequently. If you don't clean it often enough, data errors will increase and damage may be done to the tape or the recorder's mechanisms. A dirty recorder is usually a recorder that soon needs major repairs.

On the other hand, some technicians feel that too-frequent cleaning with an abrasive cleaner can prematurely wear down the heads in a VCR, so the same caution may also apply to DAT recorders, which are quite similar in mechanical design. Personally, I feel the risk from most commercially available cleaning cassettes is minimal, and certainly far less than the very real risks of dirt buildup in the delicate recorder mechanism.

TROUBLESHOOTING

In this section we will look at a few easy-to-correct problems that may occur when working with DAT recorders. In some cases, the problem may seem almost insultingly obvious when reading about it; but, believe me, it is often all too easy to completely overlook the obvious in real life. When I was working as a service technician, I was constantly amazed at how many times I'd get repair calls saying the unit wasn't doing anything at all—not even lighting up. The problem turned out to be that the cord wasn't plugged in, a fuse was blown, or, in one or two cases, a power switch was in the off position. Don't laugh. This type of mistake is far more common than you might think.

I'll share an embarrassing example that happened to me while writing this book. I use a computer word processor. I slipped in a floppy disc one morning and was unable to retrieve any of the files I'd worked on the previous day. Everything was gone. I pulled my hair and spent 15 or 20 frustrating minutes trying to retrieve the lost data until it finally occurred to me to double check the label on the disc. Sure enough, I'd slipped in a blank. My data was perfectly safe on a different disc.

The moral is, when things go wrong, take a deep breath and try to think of the simplest possible cause. Nine times out of ten, it will be something incredibly simple.

In some cases, apparent problems stem from different capabilities of different equipment. Not all DAT recorders support all features. For example, in Chapter 7, a couple of the units discussed are not designed for use with a sampling rate of 32 kHz. If a DAT cassette is recorded using this sampling rate, it cannot be played back on a unit without such capabilities. Nothing is wrong, you're just asking the machine to do something it wasn't designed to do.

From personal experience and observation, it seems clear that the vast majority of apparent problems with high-tech equipment ultimately boils down to operator error of some kind. The troubleshooting suggestions in this section will help alert you to some common operator errors.

If your DAT equipment actually needs repair, call in a qualified technician. Do not attempt to repair it yourself. Even if you've fixed analog stereos and TVs, don't open up a DAT recorder unless you've had specialized training for that type of equipment. It is very complex, with numerous delicate parts. An uninformed attempt at repair is very, very likely to result in more serious problems.

Cassette and loading problems

We'll start off with problems that might show up with the tape itself.

Problem—door to cassette well will not open Is the recorder plugged into a live socket? In an automotive installation, is the car's ignition key on to supply power to the player? Is the fuse blown?

If the recorder is getting power, is a cassette jammed in the machine? If so, do not force anything trying to get it out. If you can gently work the cassette and tape out of the recorder, great. Be sure to use a cleaning tape before using the recorder on a fresh cassette. If you cannot remove a jammed cassette easily, call a qualified service technician. Too much force can bend or break delicate tape guides and other parts in the recorder.

If the recorder is definitely getting power, and the well is empty, but the door will not open, check to see if anything is physically holding it in place. If not, call a qualified service technician. A troubleshooting chart for this problem is shown in Fig. 8-5.

Problem—cassette cannot be inserted into the well First, double check to make sure you are trying to put the cassette into the recorder rightside up. DAT cassettes can only be inserted in one direction. It simply will not fit if you try to insert it backwards. Next, check to see if there is another forgotten cassette already in the well.

On some DAT recorders and players, internal humidity or temperature sensors may prevent cassettes from being loaded into the machine under bad environmental conditions. This kind of problem is especially likely to occur in automotive installations. This is a safeguard to protect your tapes and equipment. Do not attempt to override any such protective sensors. A troubleshooting chart for this problem is shown in Fig. 8-6.

Problem—cassette is promptly ejected after it is inserted into the recorder The most likely cause of this problem is a bad tape cassette of some sort. It is not likely to be a problem within the recorder itself.

If the cassette housing is physically deformed in any way, it may not be accepted by the machine. The cassette will also be rejected if the tape is cut or broken. Most DAT recorders and players will immediately eject a cassette with condensation on the tape surface. If the cassette or tape is bad, use a different cassette. If the problem is condensation, wait for the cassette to acclimate itself to the environment. A troubleshooting chart for this problem is shown in Fig. 8-7.

Problem—automatic eject does not occur Ordinarily, at the end of a DAT cassette, the tape is automatically rewound to the beginning and ejected. If the tape is rewound, and instead of being ejected it starts playing again, the repeat function button

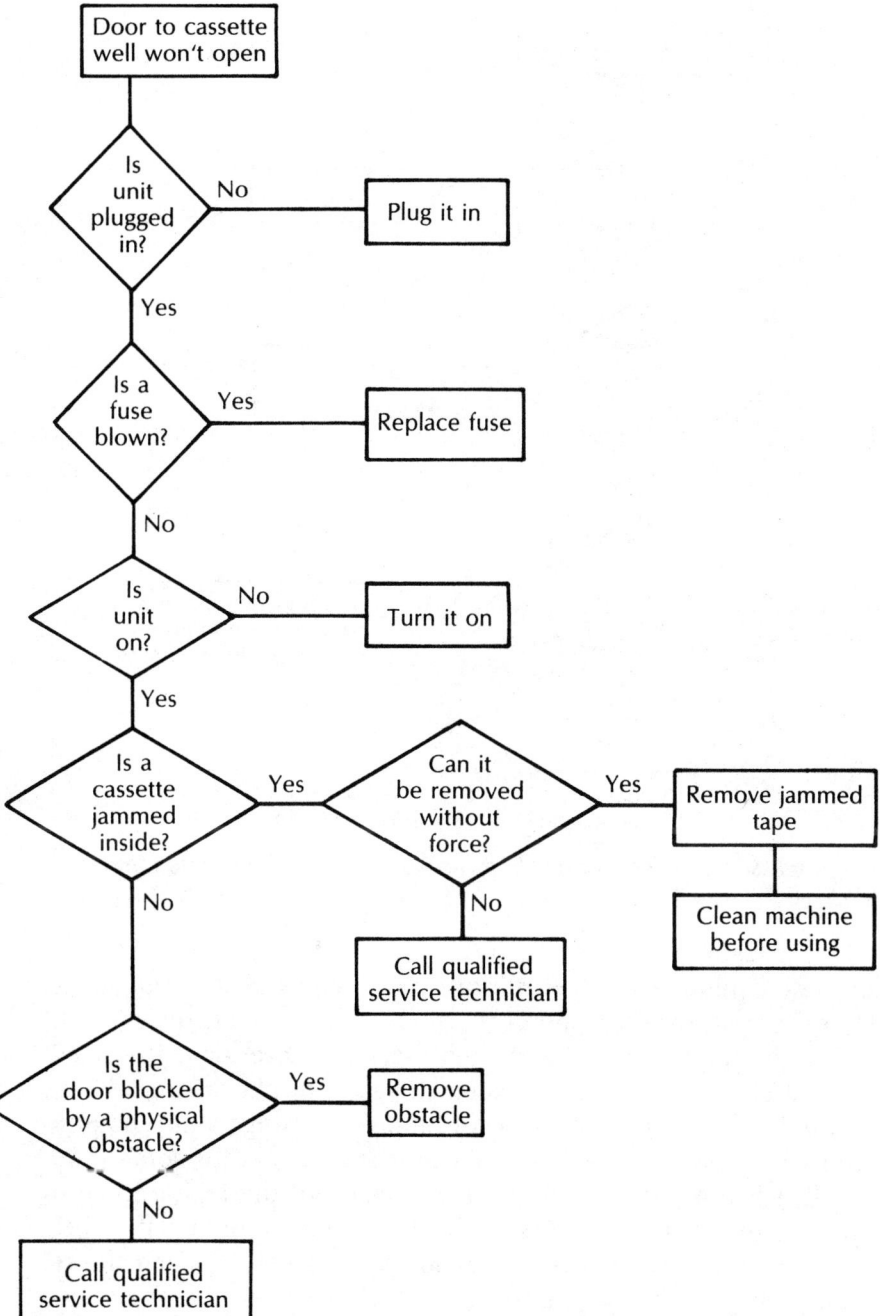

Figure 8-5 *Troubleshooting chart—door to cassette well will not open.*

216 Maintenance and Troubleshooting

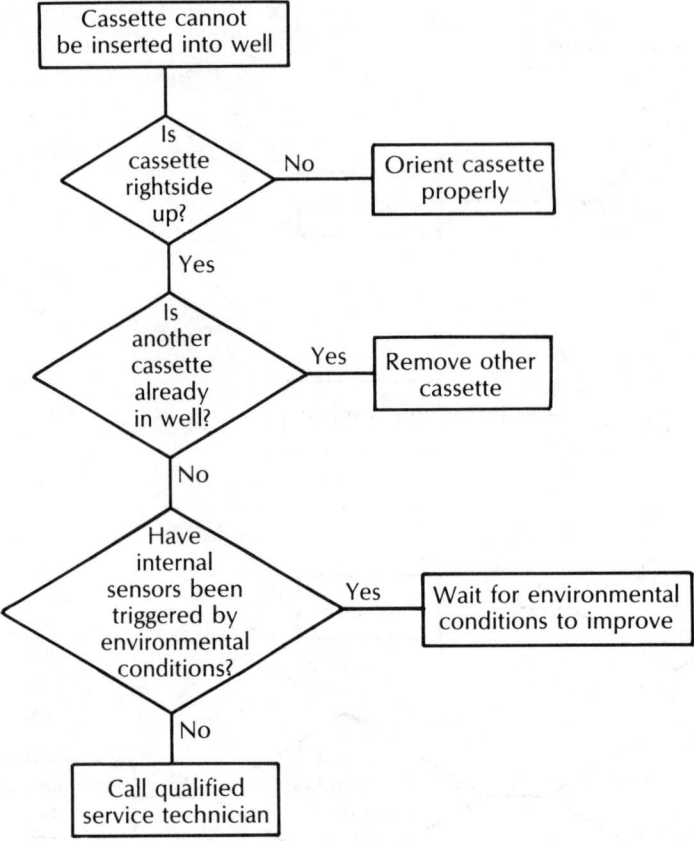

Figure 8-6 *Troubleshooting chart—cassette cannot be inserted into the well.*

has been pushed. This function may have different names on DAT recorders and players from different manufacturers.

If the tape rewinds and then just sits there, or if it doesn't rewind at all, it may be jammed. Be sure to check the owner's manual for your individual unit before jumping to any conclusions. This behavior may conceivably be normal for some models. If the tape is jammed, do not force anything trying to get it out. If you can gently work the cassette and tape out of the recorder, great. Be sure to use a cleaning tape before using the recorder on a fresh cassette. If you cannot remove a jammed cassette easily, call a qualified service technician. Too much force can easily bend or break delicate tape guides and other parts in

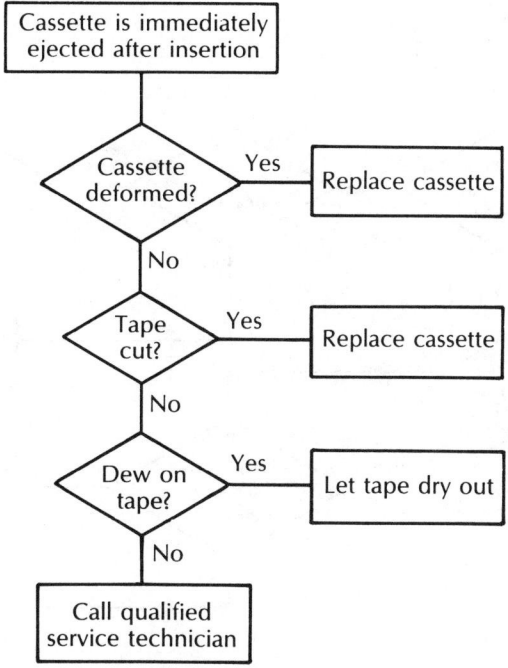

Figure 8-7 *Troubleshooting chart—cassette is promptly ejected after it is inserted into the recorder.*

the recorder, leading to a very hefty repair bill. A troubleshooting chart for this problem is shown in Fig. 8-8.

Problem—cassette is suddenly ejected during playback A change in environmental conditions (especially humidity or temperature) may have occurred, triggering the machine's protective sensors. This kind of problem is especially likely to occur in an automotive installation. Do not attempt to defeat the functioning of these sensors. They are there to protect your tapes and equipment from possible serious damage.

Alternatively, the problem may be a deformed cassette. Some physical deformities in the cassette housing will permit the tape to be inserted and played up to a certain point, then the recorder senses unacceptable tape tension and rejects the cassette. If you have a deformed cassette, replace it with a new tape. A related problem is that the tape may be improperly wound within the cassette, causing the machine to interrupt playback and eject the cassette. Such improper winding is almost always

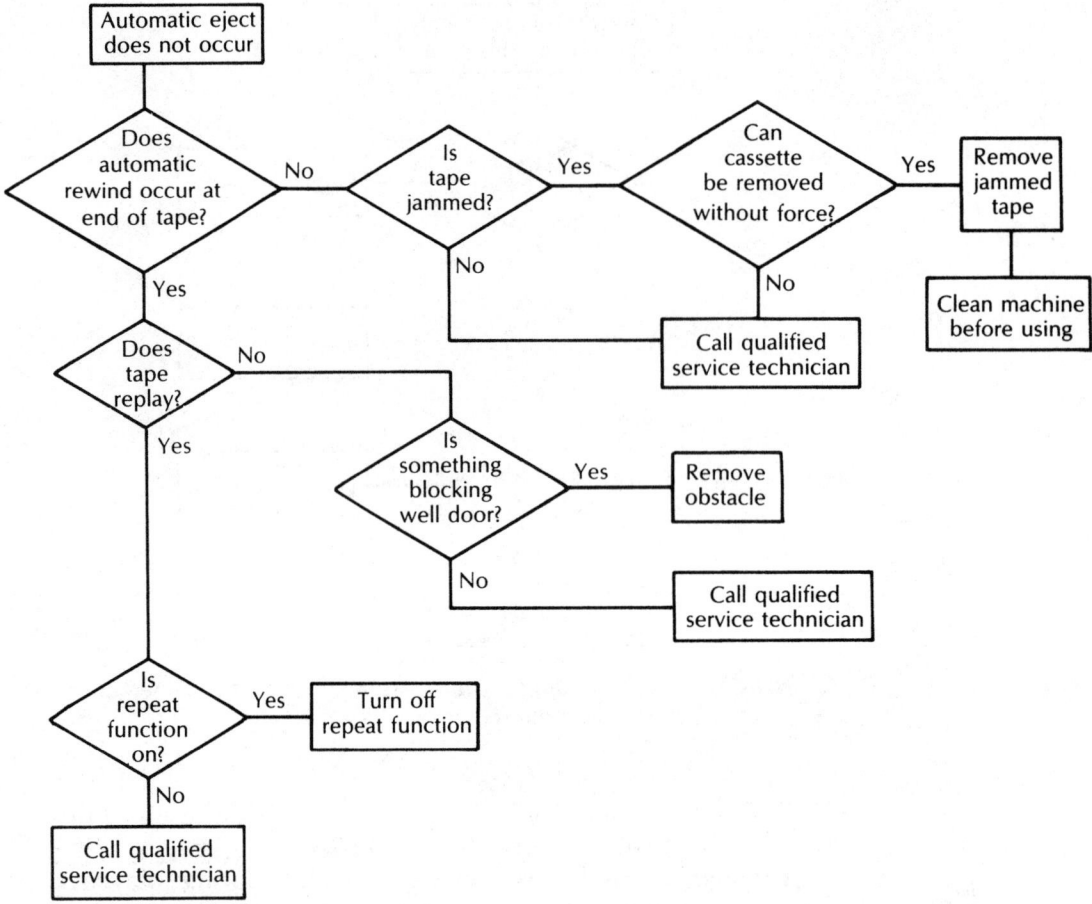

Figure 8-8 Troubleshooting chart—automatic eject does not occur.

due to some defect in the cassette housing or the tape itself. Replace the bad cassette.

On some DAT recorders, a brief interruption of the power supply voltage can cause the cassette to be ejected when the power returns a fraction of a second later. In some automotive installations, starting the car's engine may cause the player to eject a tape during playback. In either case, it's just a glitch, not a problem. Just reinsert the tape and hit the play button again. If the cassette is again rejected, assume there is something wrong with the cassette. Try a different tape. If the second cassette works, the first cassette was defective—discard it. If the second cassette is also prematurely ejected, there is apparently something wrong with your machine. Call a qualified service technician. A troubleshooting chart for this problem is shown in Fig. 8-9.

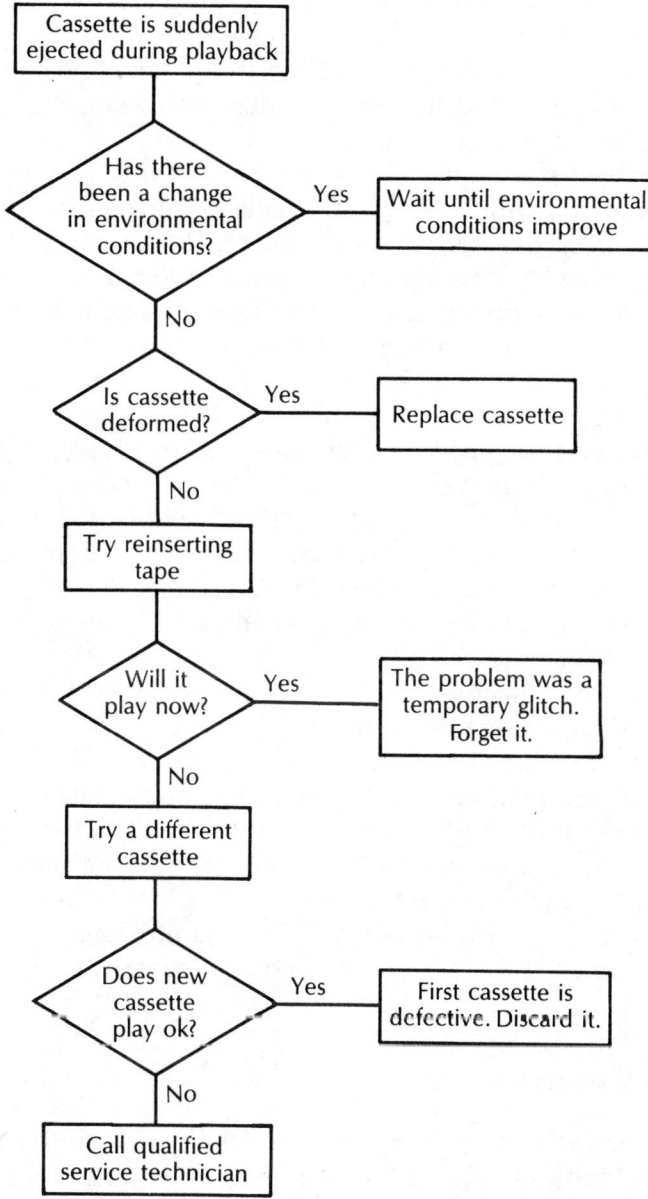

Figure 8-9 *Troubleshooting chart—cassette is suddenly ejected during playback.*

Problem—cassette is suddenly ejected during fast forwarding or rewinding The most probable cause for this particular problem is an improperly wound tape. In many cases, such improper winding is due to some defect in the cassette housing or the tape itself. Replace the bad cassette. Unlike playback ejection, however, improper tape winding problems during fast forwarding or rewinding may be a temporary glitch. Try playing the tape to the end and rewinding. If the problem does not recur, don't worry about it. It's fixed.

However, if the same thing keeps happening with the same cassette, or if it won't allow you to play to the end or rewind to the beginning, the cassette is defective. Replace it. A troubleshooting chart for this problem is shown in Fig. 8-10.

Problem—during playback, the tape is suddenly fast forwarded, then playback resumes There are two possible causes for this seemingly bizarre behavior in a DAT machine. Both stem from the way the tape was recorded.

Many (but not all) DAT recorders feature a "skip" function. A recorded program with a "skip ID" in its subcode data will not be played back when the skip function is activated. Instead, the player will automatically fast forward past the "skip" program and start playing the next selection on the tape. The "skip" function can usually be overridden by pushing a button on the player's control panel.

Most DAT machines will also automatically fast forward past any completely unrecorded section of tape. An unrecorded blank must be distinguished from a recorded blank. A recorded blank is a section of tape where there is no audio data signal, but the subcode and ATF fields are present. The PCM field is filled with zeroes, so no sound is heard. A nonrecorded blank, however, has no digital signal recorded on it. The machine does not know what to do with the tape without a control signal, so it fast forwards until it finds a suitable control signal to tell it what to do.

Functional problems

In this section we will look at some of the typical problems that can occur with the operations or special functions of a DAT recorder.

If you have a problem with a DAT machine you are not fa-

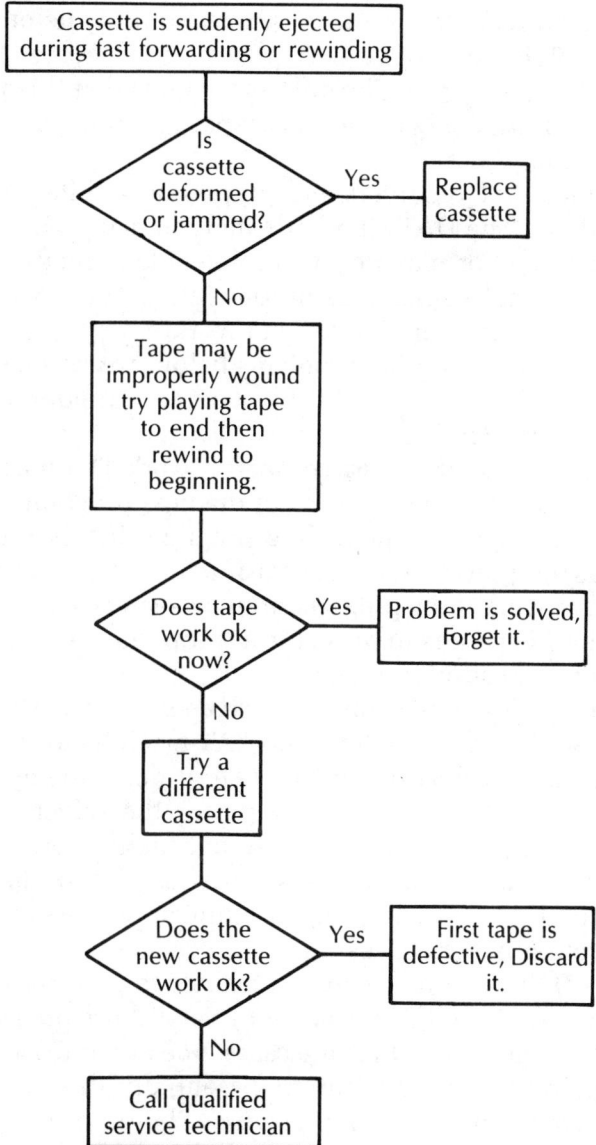

Figure 8-10 *Troubleshooting chart—cassette is suddenly ejected during fast forwarding or rewinding.*

miliar with, it is more than likely that the function you are trying to perform is implemented differently in this machine than in the model you are more familiar with. It is entirely possible that you are trying to implement a function that is not supported by the design of the machine you are using. For example, if you try to play a tape recorded at the 32-kHz sampling rate on the Kenwood KDT-99R automotive DAT player (see Chapter 7), it won't work. Nothing is wrong. The only problem is that this particular machine was not designed to play tapes recorded at the 32-kHz sampling rate.

Different DAT recorders and players have different features and functions. The DAT standards include many optional functions which may or may not be implemented on any given machine. Always take such possibilities into account when troubleshooting any function that doesn't work.

In the troubleshooting problems below, we will concentrate on other causes. The possibility of an unsupported function should be considered in each case.

Problem—next track search doesn't work The machine will not fast forward to the beginning of the next program, or rewind to the beginning of the preceding program. If this occurs, it is likely that the player's track repeat function is on. Track search also will not work if the program number search is activated. The solution in this case is to press the appropriate control button to turn off the undesired function.

Problem—program number search doesn't work The program number search function allows the DAT machine to locate a specific program or selection on the tape. It does this by checking the subcode data fields on the tape for the selected program number. Some DAT recordings may not include program numbers in their subcode data. Not surprisingly, such tapes cannot be searched for a specific program number because there aren't any program numbers to search.

Some DAT machines may not be able to perform search operations on tapes with one or more very short programs (less than about 20 seconds). In some cases, one or more start IDs may be written in too short a duration. Essentially, this means somebody goofed during recording. The tape should be rerecorded, if possible. There is nothing wrong with the playback equipment.

Problem—search operations take longer than normal Usually, a DAT machine can search for and find a specified program in just a few seconds. Under some conditions, it may take longer for the machine to locate the desired program. If the tape includes unrecorded blank sections, the search operations will be slowed noticeably.

On most tapes, the program numbers will probably be in

numerical order, which will naturally simplify and shorten the search operations. Some tapes, however, may contain program numbers that are out of sequence. It will take longer for the machine to search for and locate a specific program with such a tape.

Problem—no data display and no sound Is the machine plugged in? Is it switched on? Is there a blown fuse or circuit breaker? If the machine isn't doing anything at all, the first thing to suspect is that it is not getting power for some reason. Check both the fuses in the equipment and your power line fuses or circuit breakers. A troubleshooting chart for this problem is shown in Fig. 8-11.

Problem—no data display, but playback sound is heard Subcode data recorded on the tape may be of a type or in a form that the particular machine cannot display. If this problem shows up on all tapes, there is probably something wrong with the display circuitry. Call a qualified service technician. A troubleshooting chart for this problem is shown in Fig. 8-12.

Sound problems

The final category of problems we will troubleshoot in this chapter involves sound quality. Normally, a DAT machine provides excellent sound reproduction. If the sound is not right, something is wrong.

The first thing to check is the rest of the audio chain—the amplifier, preamplifier, and speakers. Could they be causing distortion? Is something in the audio chain disconnected? Is something connected incorrectly? Are any control settings incorrect? Bad sound is rarely a problem of the DAT machine itself, although it can happen.

Problem—no sound Is the amplifier and preamplifier on? Is there a blown fuse to one of these units? Are the speakers connected? Is the amplifier set for headphone listening?

If your amplifier, preamplifier, and speakers are working, it is possible that there is no signal on the tape you are trying to play. It could be a recorded blank (subcode data present) or an unrecorded blank (no signal recorded at all).

A tape recorded with a sampling rate not supported by the machine being used for playback will not play. It may feed past the heads, but no audio signal will be reproduced because it is

224 Maintenance and Troubleshooting

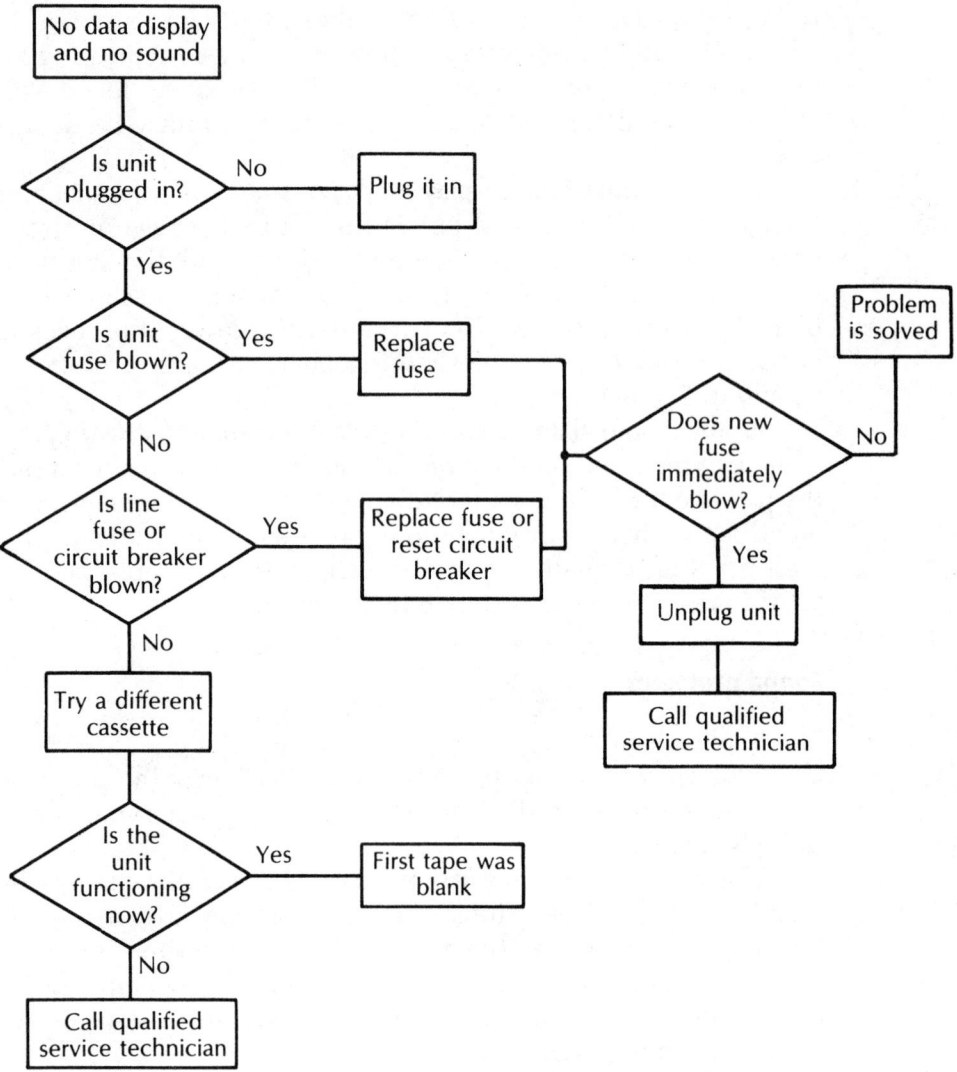

Figure 8-11 Troubleshooting chart—no data display and no sound.

in a form that is unrecognizable by the player's circuitry. This is most likely to be a problem with tapes recorded with the optional 32-kHz sampling rate. The 48-kHz sampling rate is mandatory and must be supported on all DAT recorders and players. The 44.1-kHz sampling rate is used for commercial prerecorded tapes, so it will almost certainly be supported on all DAT ma-

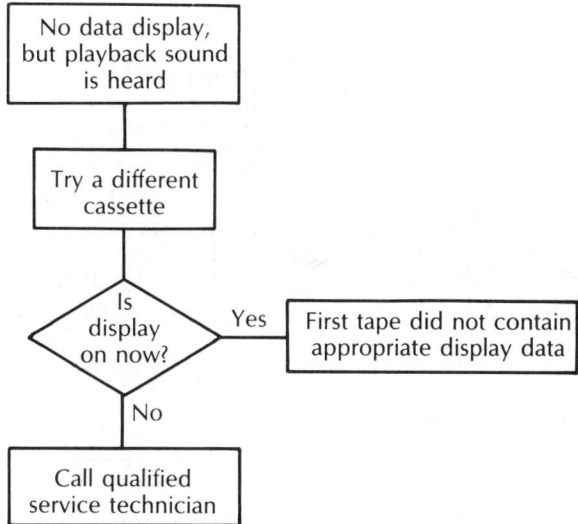

Figure 8-12 *Troubleshooting chart—no data display, but playback sound is heard.*

chines. The 32-kHz sampling rate, however, may or may not be supported. A troubleshooting chart for this problem is shown in Fig. 8-13.

Problem—distorted sound Suspect your sound system's amplifier, preamplifier, or speakers first. If the distorted sound is definitely coming from the DAT deck, it is almost certainly a problem with the tape. If the recording level was set too high, severe clipping of the signal can occur, resulting in some very unpleasant distortion. Distorted playback sound can also be the result of degraded tape. If the tape has been physically degraded, it should be thrown out. A troubleshooting chart for this problem is shown in Fig. 8-14.

Problem—crunching noise is heard This could happen if the machine's heads are dirty. Clean the heads with a cleaning cassette, as described earlier in this chapter. A crunching sound could also be the result of serious scratches or other degradation of the tape. Discard and replace the cassette in question. Of course this is assuming that you are not listening to some very avant-garde tape where a crunching sound has intentionally been recorded. A troubleshooting chart for this problem is shown in Fig. 8-15.

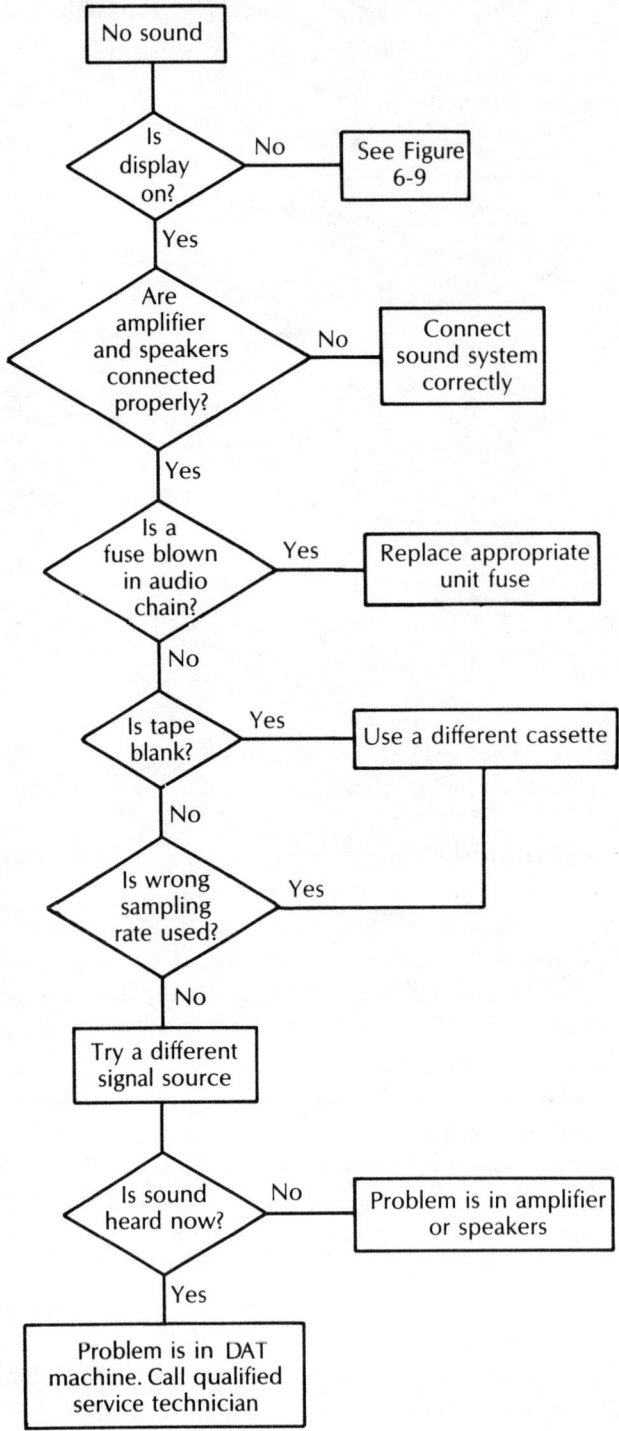

Figure 8-13 *Troubleshooting chart—no sound.*

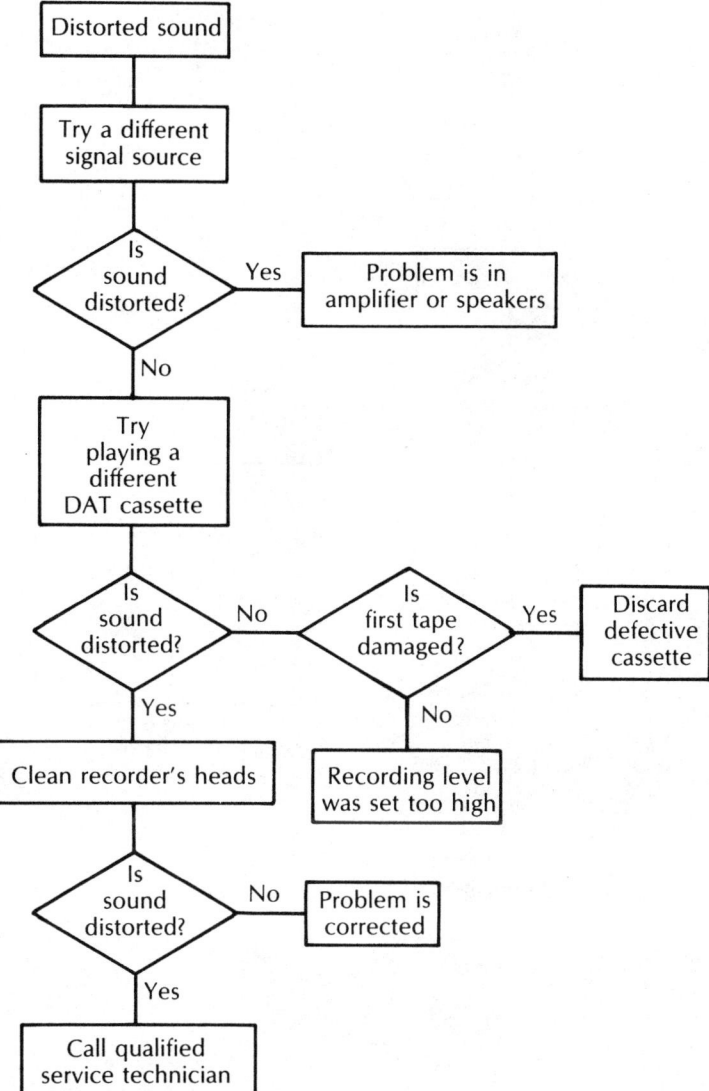

Figure 8-14 *Troubleshooting chart—distorted sound.*

If you run into any other problems that are not described here, try to apply the same logical procedure we've used here. Troubleshooting is really just a matter of applied common sense. If you cannot determine the cause of a problem, or if there is definitely a problem within the machine, call a qualified service technician.

228 Maintenance and Troubleshooting

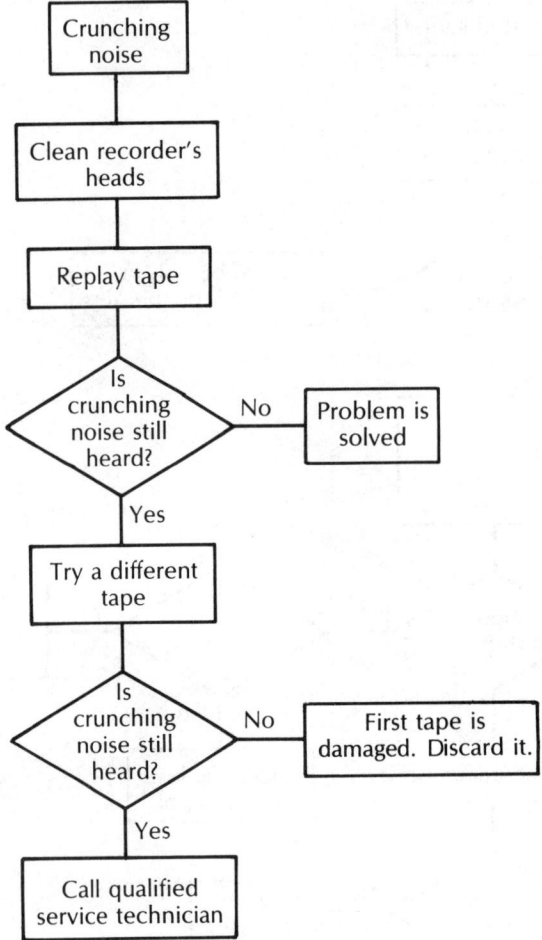

Figure 8-15 *Troubleshooting chart—crunching noise is heard.*

EMERGENCIES

If any liquid is ever spilled into a DAT recorder or player, do not attempt to use it. Unplug the unit, and call a qualified service technician as soon as possible. Some liquids (such as cola) do more damage the longer they're allowed to remain in the machine.

If you should ever see or smell smoke from a DAT machine, immediately turn it off and unplug it. Do not wait to see if the problem goes away. It could be a major fire hazard. Call a qualified service technician as soon as possible.

If a DAT machine is dropped, check it visually before attempting to use it. If it looks OK, plug it in and try it out, but be prepared to yank the plug immediately if there is any smoke, sparks, or flames. If the machine is visually damaged, do not attempt to use it. Play it safe and call a qualified service technician. It could be a shock or fire hazard. It may also be damaged in such a way that it will "eat" and permanently damage your tapes if you try to play them.

When in doubt, call a qualified service technician. Don't take chances. It could cost you more money to ignore warning signs, and in some cases it could be dangerous.

Of course, prevention is your best bet. Don't allow liquids near your DAT machine (or any other electronic equipment, for that matter). Transport it carefully. Be careful not to drop it, especially onto hard surfaces. With a little common sense and reasonable maintenance procedures, as discussed throughout this chapter, you should have few, if any, significant problems with your DAT machine.

9 ❖
The Future of DAT

JUST WHAT DOES THE FUTURE HOLD FOR THE DAT FORMAT? WE are in an area of pure speculation here, of course. In this chapter we will look at the advantages and disadvantages of this new format and the potential obstacles it faces.

WHO'S GOING TO BUY IT?

Initially, at least, the primary market for DAT recorders is surely going to be concentrated mostly in the professional and semiprofessional area. Several professional recording studios have been using digital recorders for a number of years now. The DAT standards permit greater consistency between equipment brands and individual studios, and increased convenience. The DAT cassettes are small, easy to use (and hard to misuse), and easy to store. Earlier digital recorders used bulky video cassettes or wide open-reel tape. Also, professional-quality DAT recorders are likely to be more affordable to smaller studios than previous digital recorders, or even comparable quality multitrack analog equipment.

Digital recording is highly desirable in a professional or semiprofessional recording studio for many reasons. A digital recording typically has very low levels of audible distortion, and is totally free from such annoyingly undesirable analog effects as wow and flutter. The recorded digital data can be easily manipulated in almost any way with a computer. For example, the pitch of a sound can be changed without altering the speed, or vice versa. This is a difficult and rather expensive trick using analog equipment. Many other unique special effects are also possible with digital recording.

Unlike analog recordings, digital recordings are not in any way degraded by multigeneration copies. This is a big advantage for complex multipart recordings where some parts are recorded and then repeatedly overdubbed by additional parts at later times. The recording can go through as many mixdown generations as the engineer likes and the final result is every bit as clear and noise-free as the original recording. Distortion and tape hiss does not increase with each new generation of tape as it does with analog tape recorders.

Because the DAT cassette is so small, very portable DAT recorders are a very real possibility. Portable analog recorders are usually rather bulky or have severely limited sound quality. A portable DAT recorder would be very useful for any type of location recording. A DAT recorder could be employed, for example, to capture excellent recordings of ethnic folk music in its natural habitat. In the past, recordings of this type certainly have not enjoyed the best possible fidelity. Portable DAT equipment could also improve the quality and convenience of concert recordings. A good portable DAT recorder would also be extremely useful for filmmakers who want to create the cleanest possible sound tracks for their films.

Radio stations are also likely to be prime customers for DAT equipment, especially of the portable type. A portable recorder could be conveniently used for news, documentaries, and "man-in-the-street" interviews—all of which normally require several generations of editing copies, significantly reducing the overall sonic fidelity of the final air tape. Pop stations are also likely to use portable DAT players to supply music for live remotes and special events. The DAT equipment will probably be more reliable and easier to move and set up than traditional analog sound equipment.

On the semiprofessional level, a DAT recorder is ideal for musicians and composers to make sophisticated demo tapes. The clean, distortion-free sound will surely make a good impression, and allow their music to be heard better without unnecessary technical distractions. Electronic "one man bands" who play all the parts themselves will be able to pass their tapes through as many generations as they need to create elaborate pseudo-orchestral recordings, without worrying about noise and distortion buildup.

It will also be very convenient for the musician or composer to make as many copies of his finished demo tapes as he needs, and each copy will sound every bit as good as the original master tape. In fact, the perfect copies possible with a DAT recorder may even permit some musicians and groups to form their own small record companies, manufacturing and selling their tapes themselves without going through a major record company.

Of course, this bears no relation to the piracy issue (see Chapter 6), but I'm sure the RIAA secretly hates this possibility. The existing record companies are not likely to appreciate any new competition. An independent recording company using DAT in this way is likely to face some difficulty marketing its products through existing distributors. Most independent label tapes of this type will probably be sold mostly at concerts, on consignment in local stores, and through mail order.

But what about home use? Many audiophiles have been impatiently waiting ever since the DAT format was first announced. The high sound quality offered by DAT is unquestionably a dedicated audiophile's dream come true.

Realistically, however, the dream will come true only for very well-heeled audiophiles, at least initially. DAT recorders (and related equipment) at this stage are extremely expensive. Any new product will tend to be somewhat overpriced at first. For the first couple of years of a new product's life, the consumer is, in effect, paying off the accumulated research and development costs for developing the product. New types of equipment are usually manufactured in small quantities until the manufacturers have determined just what the actual market for the product will be. It is always cheaper to mass produce any product. So the early consumer will also be paying more to compensate for the initial small manufacturing quantities, as well as retooling the factories to produce the product. The announced DAT recorders and players I've seen so far all have list prices well over $1000. Most seem to fall in the $1500 to $2500 range.

If the DAT format becomes popular, prices will certainly come down quite a bit. The manufacturer's costs per unit will drop with mass production, and much (though probably nowhere near all) of these savings will eventually be passed along to the consumer. This is not out of any particular good-heartedness of the manufacturers, who are, of course, in the business to make money. But with a popular product, manufacturers will tend to cut their prices to compete with one another on the open market. This is exactly what happened with home computers and VCRs. When VCRs first came out, $1000 would buy you a simple "bare bones" unit. Now, it's common to see VCRs on sale for $130 to $150; and these budget models are every bit as good as the early $1000 models. Old designs have been improved, and bugs have been ironed out. Low-end VCRs today typically have more features than the original, more expensive units. For example, it is virtually impossible to find any VCR today at any price level without a wireless remote control.

So, if DAT recorders catch on, the prices will surely plummet in the next few years. But that's a big "if." Will enough consumers consider the advantages of DAT worth the high initial prices to support the format long enough for manufacturers to lower the prices and make DAT more appealing to the masses? How many people will decide to play smart and wait for the prices to come down before buying? If enough people who want a DAT recorder wait for lower prices, the format may be killed off because the initial sales are insufficient.

This is related to what happened with quadrophonic sound, discussed back in Chapter 1. Of course, in that case, the public was waiting to see which of several formats would dominate. The reasons for the delayed sales are irrelevant, however. If a sane manufacturer doesn't sell enough units, he's definitely not going to step up production, and it's the stepped up production that will lead to lower wholesale and retail prices.

Besides the high sound quality, DAT's biggest selling point to the general public is likely to be the very long recording times available in this format. Up to 2 hours can be recorded on a standard DAT cassette (4 hours on the slow speed). This is considerably longer than a CD can hold. The CD format has made a lot of headway in the marketplace with its long playing times (even though this time advantage hasn't been taken advantage of with all current commercial releases). DAT offers the further advantage over CDs of being user recordable and erasable.

Many industry observers believe longer recording times was the chief advantage of VHS VCRs over Beta VCRs. The Beta format was considered to be slightly better technically than VHS, but VHS offered somewhat longer recording times; and Beta is virtually dead now. Good tape economy seems to appeal to much of the general public (though not necessarily to the true audiophile).

In practical terms, it is a good thing that DAT cassettes have such long recording times, because they will be relatively expensive, and the costs in this case are not likely to come down significantly. The cassette housing has many delicate parts which must be precisely assembled. Also, the tape itself will continue to be rather expensive. Ferric oxide tape, used in most analog cassettes and video cassettes, is relatively cheap, but the raw materials needed to make metal particle tapes (required by the DAT format) tend to be quite expensive.

Will audiophiles be able to support the DAT format until it catches on with the general public? Will DAT catch on with the general public at all, or will it be seen as an unnecessary frill or just one more gimmicky gadget? The big question is just how attractive will the DAT format be to the general (nonaudiophile) population? This will probably depend primarily on the availability of popular commercial recordings and the public's perception of the competition from other available formats. These factors will be discussed in the next few sections of this chapter.

CONTINUED OPPOSITION FROM THE RIAA

There is little room for question that the major record companies and the RIAA are less than thrilled with the whole idea of the DAT format. They are not likely to offer much support to the format unless it really takes off as an obvious profit maker.

We have already looked at the legal problems (see Chapter 6). The SCMS copy protection system reduces, but does not entirely eliminate, the objections of the RIAA. It is possible that some record manufacturer may yet file a copyright infringement lawsuit against a DAT manufacturer or user.

This is exactly what happened with the Beta VCR. A couple of movie companies sued the Sony Corporation and one of their customers for copyright infringement. Eventually, the courts decided in favor of Sony and their poor customer. The individual Beta user who was named in the lawsuit was just picked more or less at random. To be valid legally, the lawsuit had to name some specific individual. To Sony's credit, they paid all the legal expenses of the case, even though they were not legally responsible for their customer's defense.

Even though Sony finally won the case, it was a long, drawn-out and expensive fight, and it could have conceivably gone the other way. SCMS reduces the odds of such a lawsuit, but does not entirely eliminate the possibility.

Aside from the highly controversial copying issue, many of the larger record companies don't particularly like the idea of entering yet another recording format into the market at this time. LPs are currently being phased out, but analog cassettes and CDs are both selling quite well. Several of the major record companies think the new DAT format would confuse consumers, splitting up existing sales and forcing them to retool their factories for a new format, without any additional profitability.

It is very unlikely that we'll be seeing many prerecorded DAT cassettes for the next few years, at least not from any of the major labels. This just may give the extra needed toehold for a number of newly formed independent labels. Starting up a new label on DAT would be far less expensive and more practical than starting a new label for CDs, LPs, or even analog cassettes. No other present day recording medium offers a more convenient approach to extremely high-quality tape duplication than DAT. Mass producing duplicate copies without degrading the sound quality is no problem within the digital realm.

Early prerecorded DAT releases are likely to be heavy on classical selections. This prediction is based on two ideas. One, audiophiles, who place the most emphasis on realistic sound reproduction, often tend to be into classical music. Naturally, audiophiles can be expected to be the most likely early customers for DAT machines, so their tastes are likely to be influential. Orchestral and choral recordings with their widely varying dynamic and tonal ranges are generally the most critical, and so any reproduction defects are generally much more noticeable with recordings of this type of demanding music. Since, initially, DAT recorders are going to be bought mainly by serious audiophiles, the biggest market for early DAT cassettes can be expected to be strongly skewed towards the classical area.

A second reason why we'll probably see a lot of classical DAT releases early on, is that the length of a standard DAT cassette is simply so well suited to major classical pieces. Entire symphonies, and even the shorter full-length operas, will fit neatly and uninterrupted onto a single cassette. It would take a lot of 3-minute pop songs to fill up a 2-hour tape. For popular music, a royalty must be paid by the record producer for each copy of each selection recorded. On a 30- to 40-song compilation, this could really add up.

When the major record labels start to release prerecorded DAT cassettes, don't expect them to routinely fill up each tape's time capacity. Even with CDs, which can hold up to 74 minutes, most current CD releases have only about 35 to 50 minutes of music. Many CDs have less than 30 minutes of music. More than half of the storage capacity of the CD is being wasted on these releases. Prerecorded DAT cassettes from major labels will probably be made on customized cassettes with less tape than the 2-hour standard.

The initial scarcity of prerecorded DAT releases is likely to continue for quite some time. This will almost certainly slow down sales for the format, at least to some degree. Except for people into recording for its own sake, most consumers seem to prefer buying ready-made tapes. If there isn't a good selection of commercial recordings, will people bother to buy DAT recorders and players at all. The shortage of commercial releases in this format could conceivably cripple or even kill DAT before it really has a chance to blossom. Naturally, I hope that doesn't happen, but I think we need to consider the realistic possibilities here.

THE COMPETITION FROM LPS

Ultimately, the success or failure of the DAT format will depend on how the buying public sees it in comparison to the existing competing formats. At the present time, there are three active music reproduction formats for consumers—the LP, the analog cassette, and the CD. We will compare and contrast each of these existing formats with DAT in the following pages.

The LP has been around for a long time. It is, by far, the oldest existing sound reproduction format. It's chief asset is its familiarity. LP discs are a playback-only medium. The consumer cannot make his own recordings in this format. DAT, of course, is a record and playback medium.

Because the LP has been around so long, there are a great many existing recordings in this format. Many consumers already have fairly extensive collections of LPs. Many earlier recordings are available only in this format and in many cases are unlikely to ever be rereleased in a more modern format.

As we've already discussed, prerecorded DAT cassettes are virtually nonexistent at present, and are very likely to remain scarce for several years to come. On the other hand, the LP is unquestionably dying out. A number of recent commercial releases have been in the analog cassette and CD formats only, with no LP version.

A record turntable permits a limited form of random access to any point in the recording. The tone arm can be manually positioned over the approximate spot where the user wishes recording to begin. A DAT recorder, like any tape recorder, is sequential in nature and cannot offer true random access. However, the digital search functions using program numbers in the subcode data can be very fast. In the search mode, the tape runs at speeds up to 150 to 200 times the normal playing speed. In effect, the search function will seem almost like random access. It can certainly be more accurate than physically positioning a turntable's tonearm.

A record disc with its physical grooves can be easily damaged. Scratches are very hard to avoid. The disc itself can be cracked or broken. The very process of playing a record, with the stylus riding in the groove, damages the record, eating away the groove walls. This is especially true if a high-pressure ceramic cartridge is used (see Chapter 2).

A DAT cassette is much more durable, especially since it is completely contained in a hermetically sealed housing. Damage is possible, however, and the physical contact between the tape and the recorder's heads and tape guides does put a certain amount of wear and tear on the tape. Jamming can be a problem with any tape format, especially when a cassette housing is used.

A tape, such as a DAT cassette, can be accidentally erased. This is not possible with an LP, of course.

All in all, the LP is not likely to present major competition to the DAT format. The old vinyl discs are already rapidly disappearing from dealer's shelves. There are still a few diehards earnestly defending the LP, and even insisting on its superiority. I suspect this reaction is largely due to a reluctance to replace large existing collections, and general technophobia. Some people feel very threatened by anything new. They seem to feel accepting a new way of doing things would somehow invalidate the way they did things before. Yesterday's best may be surpassed today.

The LP turntable is simple and elegant, and its familiarity is reassuring. You can watch the record spinning around and see the tonearm over the spinning record. You can even watch the stylus riding in the groove to reproduce the recorded sound. In a digital system, such as DATs and CDs, it all seems so complex and mysterious. Playback operations take place behind closed doors. The tape or disc is placed completely inside the machine. Even if we could see inside, there wouldn't be much of anything to see—most of the "action" is purely electronic in nature. It seems to be part of human nature to fear (or at least mistrust) what one doesn't understand, and the good old LP is certainly far easier to understand than digital electronics.

But this is the computer age, and more and more general consumers are becoming at ease with high-tech gadgets. The LP proponents are in a definite minority, regardless of the merits of any of their arguments. It is doubtful that any vinyl LPs will be manufactured (at least by any of the major labels) after about 1995.

THE COMPETITION FROM ANALOG CASSETTES

The CD was in direct competition with the LP. It is essentially a digital version of the old vinyl disc. Both are playback-only media. Similarly, DAT recorders are most directly in competition with analog audio cassette recorders. Both are magnetic tape record/playback systems, with the tape enclosed in a small, convenient cassette housing. Their primary intended applications are directly parallel.

One of the strong selling points of analog cassettes is their small size. In fact, Phillips, who invented this format, called it the "compact cassette." DAT cassettes have an edge here, since they are about half the size of the standard analog audio cassette (see Table 9-1).

Table 9–1 Dimension comparisons between analog and DAT cassettes.

Dimension	Analog Cassette	DAT Cassette
Length	102.4 mm	73 mm
Width	63 mm	54 mm
Depth	10.5 mm	12 mm

As you can see, the biggest difference is in the length of the two types of cassettes.

On the other hand, it is likely that unsophisticated consumers may be somewhat put off by the one-sided nature of the DAT format. They may feel that they are being "cheated" because they cannot turn a DAT cassette over and use the other side, as they can with an analog cassette. Of course, the DAT format uses the entire width of the tape, so there can be no Side 2.

Judging from the history of VCRs, longer recording times seem to have strong sales value. The DAT has an advantage here. The standard DAT cassette can hold up to 2 hours at the normal recording speed, without turning the cassette over.

The maximum length for an analog cassette is the C120, which can record up to 1 hour on each side; for a total of 2 hours. However, C120s are rarely used, because they tend to be unreliable. The tape in C120 cassettes is very thin, and is easy to stretch or break. The C120 cassettes are also more prone to breaking. Most analog cassettes are C60s (30 minutes to a side) or C90s (45 minutes to a side).

The DAT's time advantage is even greater than this. On the slow speed, supported on some (not all) DAT recorders, a standard DAT cassette can hold up to 4 hours of music. Extended length (thinner tape) DAT cassettes will also be available. These extra long cassettes will hold up to 3 hours at the standard speed, and up to 6 hours at the optional slow speed. Clearly, the DAT format wins the time battle, hands down.

The other major selling factor is price. DAT cassettes will certainly be considerably more expensive than analog cassettes. The housing of a DAT cassette is more complex and more thoroughly sealed than an ordinary analog cassette. Of course, this will add to the manufacturing costs. Moreover, most analog cassettes use ferric oxide tapes, whereas a DAT recorder requires metal-particle tape, which is more expensive. Metal-particle analog cassettes are available, but they don't sell nearly as well as their cheaper ferric oxide counterparts.

For the following technical specifications, we will assume that metal-particle tape is being used in the analog cassette recorder, as well as in the DAT recorder; this is to be as fair as possible to the analog cassette format. Several of the specifications would be degraded if ferric oxide tape was used. Because specifications vary from machine to machine, we will just be comparing typical values here.

A typical analog cassette recorder using metal particle tape can reproduce frequencies as low as 25 Hz and up to 20 kHz (20,000 Hz). This is good, but a typical DAT recorder's frequency response specs are even better. Most DAT recorders can accurately reproduce frequencies as low as 2 Hz and as high as 22 kHz (22,000 Hz).

It is debatable how significant the difference is here, because the human ear can only hear frequencies from about 20 Hz to 20 kHz (20,000 Hz). However, the frequency response for a DAT recorder tends to be much more flat across its entire range of frequencies than that for an analog cassette recorder.

The DAT format also has a significant advantage where dynamic range and signal:noise ratio are concerned. Typical values for both specifications for a DAT recorder range from about 88 to 98 dB, without any special noise reduction circuitry. A good-quality analog cassette recorder will have a dynamic range and signal:noise ratio specifications in the 50- to 60-dB range, with noise reduction circuitry on.

Some listeners find that noise reduction circuitry can color the sound somewhat. Personally, I have always found the sound of the popular Dolby B noise reduction system unpleasantly "metallic."

The THD specification for a typical analog cassette recorder is about 0.5%, which is good. But a typical DAT recorder does even better, with a typical THD specification of 0.005% or less.

Any analog tape recorder, including analog cassette recorders, is subject to minor fluctuations in the tape speed known as "wow" and "flutter." Wow is the result of low-frequency speed fluctuations, and flutter is the result of high-frequency speed fluctuations. A good-quality analog cassette recorder will have a wow and flutter specification of about 0.018%. Wow and flutter are not a problem with a digital recording medium, such as a DAT cassette. The signal pitch and speed are determined by the numerical values, not by the actual speed of the tape's movement past the playback head.

Wow and flutter are included on most specification sheets for DAT recorders, simply because most educated audio consumers are used to seeing it. The Wow and flutter specification for most DAT recorders is listed as "below measurable limit," or something similar. There may be some minute wow and flutter in the signal somewhere, but there is no way to prove it is there with modern test equipment. The effect would certainly be inaudible, in any event.

On the technical level, DAT recorders are significantly superior to analog cassette recorders. However, except for dedicated audiophiles, the general consuming public doesn't seem to place a great deal of importance on technical quality. Low-grade stereo systems continue to be the biggest sellers, and many listeners are perfectly satisfied with a technically mediocre system. An untrained ear may have difficulty hearing the improvement offered by a higher-quality sound system.

Most experts agree that in videotape, Beta was technically superior to VHS, yet VHS has become the standard video format, and Beta has virtually died out. The general (nonaudiophile) public tends to be more concerned with convenience features and cost than with technical specifications.

The DAT format is at a definite advantage when compared to analog cassettes on the cost level. But DAT can offer some unique convenience features. It is impossible to say yet whether they will be considered important enough to the U.S. consumer to make up for the cost.

Many high-end and some mid-level analog cassette recorders include automated music search functions of some sort. This feature is called different things by different manufacturers, but the acronym "AMS" is often used. This is a convenient feature. The tape is fast forwarded past the head to the beginning of the next selection (or previous selection, if rewinding) on the tape.

Analog AMS systems are relatively crude. Basically, they just look for blank spots on the tape. They can be fooled by quiet passages within a selection. The system may also run into problems if there isn't a long enough pause between selections.

A DAT recorder can do a much better job with this type of feature. The beginning of a selection can be precisely located by looking at the digital subcode data. Moreover, individual program numbers can be assigned to each selection, so any desired song can be quickly and precisely located on the tape. Index points can be added at any point on the tape, even in the middle of a passage of music.

A DAT recorder also fast forwards and rewinds much faster than an analog cassette recorder. A standard 2-hour DAT cassette can be searched from beginning to end in about 40 seconds.

A drawback to the DAT format is the severe lack of prerecorded releases. Just about every album released in the last 20 years or so is available on analog cassette. The shortage of DAT software (commercial recordings) is likely to continue for some time, and will almost certainly have an adverse effect on the sales of DAT recorders to the general public. Hopefully, it will not be enough of an ill effect to prematurely kill off this promising new recording medium.

THE COMPETITION FROM CDS

As a record/playback medium, DAT's chief competition comes from analog cassettes. As a digital medium, DAT's chief competition comes from CDs. CDs have been surprisingly successful in the marketplace, and are largely responsible for the large drop in LP sales. The advantages of CDs include

- Small, convenient size,
- Ease of use,
- Durability—CDs are close to indestructible
- Long playing times—up to 74 minutes,
- Excellent sound quality, and
- Programmability.

With a CD player, the user can have the tracks (songs) on a disc played in any desired order. Individual tracks or the entire disc may be repeated. Some CD players include a feature to play the tracks in a random order. Random access of tracks is easy to accomplish with a CD player. Any track on the disc can be accessed almost instantly.

DAT machines feature similar programmable functions, but because of the sequential nature of the tape medium, true random access isn't possible. A specific track on a DAT cassette can be found very quickly, but locating a specific track on a CD is much faster.

DAT machines use a higher data transmission speed than CD players, and there is more space for subcode data on a DAT. DAT has about 4.6 times the subcode data capacity of a CD. Much of this capacity will be unused on early equipment and tapes, but it is available to support new special functions as they are developed or needed.

Some possible uses of the extra subcode space in the DAT format include calendar information, catalog numbers, and even alphanumeric titles for individual programs and the tape as a whole.

Unlike most other sound storage media a CD is played without anything touching its recording surface. The CD is played by reflecting a laser light beam off of its surface. There is no wear and tear on the CD during ordinary playing. The millionth playback should sound every bit as good as the first. Dirt or scratches can cause significant losses of data, however, so early claims of the virtual indestructability of CDs were exaggerated. They are, however, incredibly durable. I've used over 500 CDs, including one I found lying at the side of the road with no case. So far, I've only encountered one CD that was so badly scratched it couldn't be played. (No, it wasn't the one I found in the street.)

In a DAT recorder or player, of course, the tape rubs against the head drum and tape guides during playback and recording. Strong binders in the tape used in DAT cassettes can slow down tape wear, but can't eliminate it altogether. As the tape is worn, the number of dropouts will increase. Eventually, you can expect the number of dropouts to increase beyond the capabilities of the error correction system to cover up. Substantial data loss will occur.

This is not likely to be a significant problem for most users, however. It is estimated that an average DAT cassette will be good for more than 1000 passes (either recording or playing). I don't know about you, but personally, I'd be pretty sick of the music on any tape after it had been played 1000 times.

The CD is already fairly well established on the marketplace, so it will be very stiff competition for DAT. Once again, a significant factor is likely to be the availability of commercial recordings. Thousands of titles are now available on CD. Virtually all recent releases (at least from the major record labels) have been offered on CD. Popular recordings predating the development of the CD are being rereleased in this format. Prerecorded DAT tapes, on the other hand, are likely to be quite scarce for some time to come, as mentioned several times throughout this chapter.

DAT has two main advantages over the CD that could help it in the marketplace. One, the CD is a playback-only medium, while with DAT, a user can make his own tapes, either live or compiled from other recordings. Similarly, if you end up with a tape you don't like, you can erase it and reuse it.

The second advantage of DATs over CDs is the longer running time. A CD can hold a maximum of about 75 minutes of music. At the standard speed, a standard DAT cassette can hold up to 2 hours of music. At the optional slow speed, the same cassette can record up to 4 hours. (In the "wide" playback-only mode, the standard DAT cassette holds a mere 80 minutes of music—still more than a CD.) Better, longer, DAT cassettes will be available, offering 3 hours of running time at the standard speed, and 6 hours at the optional slow speed.

It is possible that one of these advantages may be taken away before DAT gets a chance to establish itself in the marketplace. Several companies are working on creating a user-recordable CD. Recording on such a disc is likely to be less convenient than a cassette tape. Features such as "punch in" and "punch out" will probably be impossible. From what I've read, the self-recorded discs are likely to be either nonerasable, or only bulk erasable. That is, you can't erase just one song, you have to erase the whole disc and start over.

Still, these limitations may not matter very much to the general consumer, especially when they can make their own recordings that can be played back on existing CD players. If a recordable CD system is developed and it is not too expensive (two fairly big ifs), it could offer too much competition for the DAT format to survive.

Radio Shack, for example, claims to be close to having a marketable CD recorder. While they won't offer any prerelease details, they have announced that such a unit may be available in late 1990 or 1991. I read in one announcement that they were shooting for a $500 price tag.

If Radio Shack (or some other manufacturer) succeeds in this endeavor, it is very probable that DAT will become strictly a professional format. A few semiprofessional recorders may be marketed, but they are likely to be few and far between and rather expensive; comparable to the situation for analog reel-to-reel tape recorders today. On the professional/semiprofessional level, DAT has many clear-cut advantages. On the consumer level, however, it's a completely different question.

CONCLUSIONS

Will DAT be a marketing success or a flop? Only time will tell. There are too many variables involved for anyone to be too sure of themselves.

It would be a shame if DAT died out as a consumer medium, because it offers some unique advantages. But will these advantages be strong enough to lure the general consumer to support the format? Who can say? As I see it, the two biggest obstacles to home DAT are the initial high cost of the equipment and the lack of a large selection of prerecorded DAT cassettes.

What does the future hold for DAT? We'll have to wait and see. The next few years are certainly going to be interesting as this exciting new format fights to establish itself.

Appendix
DAT Manufacturers

THE FOLLOWING MANUFACTURERS ARE EITHER NOW MARKETING DAT equipment (either in the U.S. or abroad), or have indicated their intention to do so in the near future.

Aiwa
35 Oxford Dr.
Moonachie, NJ 07074

Alpine
19145 Gramercy Pl.
Torrance, CA 90501

Blaupunkt
2800 South 25th Ave.
Broadview, IL 60153

Casio
570 Mount Pleasant Ave.
Dover, NJ 07801

Clarion
661 W. Redondo Beach Blvd.
Gardena, CA 90247

Eclipse
19600 Vermont Ave.
Torrance, CA 90502

Hitachi
401 W. Artesia Blvd.
Compton, CA 90220

JVC
41 Slater Dr.
Elmwood Park, NJ 07407

Kenwood
2201 East Dominguez St.
Long Beach, CA 90801

Mitsubishi
800 Biermann Ct.
Mount Prospect, IL 60056

Nakamichi
19701 S. Vermont Ave.
Torrance, CA 90502

Sanyo
21350 Lassen St.
Chatsworth, CA 91311

Sharp Electronics
Sharp Plaza
Mahwah, NJ 07430

Sony
Sony Dr.
Park Ridge, NJ 07656

Technics
One Panasonic Way
Secaucus, NJ 07094

Toshiba
82 Totowa Rd.
Wayne, NJ 07470

Index

A

absorption of sound, 24
acoustics, 24
albums (*see* disc recordings)
aliasing, 88-90, 137
amplitude modulation (AM), 46, 85
analog cassettes, digital audio tape (DAT) vs., 238-242
analog recording basics, 49-74
analog-to-digital sound conversion, 80-82
audio cassettes, 38-41
automotive DAT decks, 180-188
  Clarion Audia 8100 am/fm DAT player, 185-187
  condensation problems, 181-182
  Kenwood KDT-99R, 182-185
  Mitusbishi dt10 DAT player, 187-188

B

bandwidth, frequency, 86
Berliner, Emile, disc records, 9
bias
  digital audio tape (DAT), 99
  tape recorders, 72-74
binary numbering system, 77-79
bits, 79
bytes, 79

C

carrier signals, digital recording basics, 84-85
Casio DA-2 DAT home recorder, 189-192
cassette care, DAT cassettes, 205-210
cassettes (*see* audio cassettes)
CD players, 123-127, 130-131
ceramic phonograph cartridges, 51-52
Clarion Audia 8100 am/fm DAT player, 185-187
cleaning DAT equipment, 210-212
compact discs (CD), 2, 47-48, 82-83, 109-134
  capacity for recording, 111, 120
  CD players, 2, 123-127, 130-131
  digital audio tape (DAT) vs., 128-131, 242-244
  digital mastering, 112-113
  distortion, 121-122, 128
  flutter, 128
  formats, 119-120
  lasers, 109, 113-118, 124-127
  manufacturing process, 132-134
  materials, 120-121, 132-134
  physical dimensions/standards, 119-120
  playback process, 122-127
  SPARS code, 131-132
  speed control, 110-111, 122
  total harmonic distortion (THD), 128
  tracking systems, 124-127
  videodiscs, 118-119
  warping, 121-122
  wow, 128
condensation problems, 181-182, 209-210
copy protection schemes, 166-179, 234-236

Index

copy protection schemes (*con't*)
  notch filters, 171-176
  SCMS copy protection, 176-178, 234-235
copying of recordings
  copy protection schemes, 166-179, 234-236
  digital recording basics, 96-97
  legality of copies, 166-178
crossover networks, 24
crosstalk, 33-34, 55

D

de-emphasis of sound, 24
decimal numbering system, 76-77
digital audio tape (DAT), 135-165
  aliasing, 137
  analog vs., 3
  analog cassettes vs., 238-242
  ATF fields, 150-151
  automotive DAT decks, 180-188
  capacity for recording, 236
  compact discs (CD) vs., 128-131, 242-244
  copy protection schemes, 166-178, 234-236
  copying, legality issues, 166-178
  disc recording competition, 236-238
  distortion, 128, 201-202
  drop-outs, data tracks, 99
  editing techniques, 137, 207
  equipment, 180-203
  error correction, 152
  fidelity of recording, lifespan, 154-155
  flutter, 128, 201
  "gray market" equipment, 178-179
  head drum, 145-147
  home DAT recorders, 188-200
  legal issues, 166-179
  lifespan of tape, 154-155
  linearity, 99
  maintenance and troubleshooting, 204-229
  market analysis, future of DAT, 230-234
  multiplexed signals, 99
  notch filter copy protection, 171-176
  PCM converters, 135
  PCM data fields, 149-152
  playback mode, normal, 163-164
  playback mode, wide, 164-165
  price comparison, 233-234
  quantization, 137
  R-DAT vs. S-DAT, 139-141
  record bias, 99
  record/play speed, option 1, 160-161
  record/play speed, option 2, 161-162
  record/play speed, option 3, 162-163
  record/play speed, standard, 160
  recorders, requirements for, 98-101
  recording process, 147-155
  RIAA opposition, 234-236
  rotating-head recorders, 99-101, 103-105
  SCMS copy protection, 176-178, 234-235
  signal sampling, 155-160
  specifications/operations, 160-165, 200-203
  specifications/operations, recorders, 141-143
  standards and formats, 138-141
  stationary-head recorders, 99, 101-103
  stereophonic recording, 153-154
  subfields, 152
  tape requirements, 98-99, 143-145
  total harmonic distortion (THD), 128, 201-202
  VCRs and PCM converters, 105-107
  wow, 128, 201
digital compact disc (see compact disc)
digital mastering, 112-113
digital recording basics, 75-108
  aliasing, 88-90, 137
  amplitude modulation (AM), 85
  analog recording vs., 75-76
  analog sound conversion to digital bits, 80-82
  carrier signals, pulse waves, 84-85
  compact discs (CD), 109-134
  copying, 96-97
  digital audio tape (DAT) recorders, requirements, 98-10
  distortion, 95-97
  drop-out, 92
  editing techniques, 108, 137
  error correction codes, 92-96
  filtering, 90

Index *251*

frequency bandwidth, 86
frequency shift keying (FSK), 84
linearity, 87
multiplexed signals, 98
numbering systems, 76-80
pros and cons, 107-108, 136
pulse amplitude modulation (PAM), 85
pulse code modulation (PCM), 86-87, 98
pulse number modulation (PNM), 86
pulse position modulation (PPM), 86
pulse width modulation (PWM), 85-86
quantization, 91-92, 107, 137
resolution, 90-92
sampling frequency, 87-90
signal-to-noise ratio, 86, 91
signals, 82-87
specifications, 200-203
direct-to-disc recordings, 111-113
disc recordings, 3, 9-11, 49-56
   crosstalk, 55
   digital audio tape (DAT) vs., 236-238
   digital mastering, 112-113
   direct-to-disc recordings, 111-113
   editing techniques, 16
   equalization, 55
   flutter, 55, 56
   groove cutting/tracing, 54-55
   history and development, 9-11
   materials used in discs, 16-17
   speed (rpm) standardization, 12-16
   stereo process, 54-55
   tape-to-disc recording, 21-22
   wow, 55, 56
distortion, 24, 225-228
   compact discs (CD), 121-122, 128
   digital audio tape (DAT), 201-202
   digital recordings, 95-97
drop-out, 92
   digital audio tape (DAT), 99
droppage, DAT equipment damage, 229

E

Edison's cylinder phonograph, 3-7
editing techniques
   digital audio tape (DAT), 137, 207
   digital recordings, 108, 137

direct-to-disc recordings, 16
tape recordings, 19-21
eight-track tape cartridges, 36-38
ejection problems, DAT equipment, 213-221
equalization
   disc recordings, 55
   tape recordings, 68-72
erase heads, tape recorders, 60
error correction codes, digital recordings, 92-96

F

filtering, digital recordings, 24, 90
flutter
   compact discs (CD), 128
   digital audio tape (DAT), 201
   disc recordings, 55, 56
   tape recordings, 31, 74
flux, 67-68
frequency bandwidth, 86
frequency modulation (FM), 46
frequency response, 61-66
   speakers, 25, 62
   tape recordings, 61-66
frequency shift keying (FSK), 84

G

gramophone (*see* disc recordings; phonograph)
"gray market" DAT equipment, 178-179

H

heads, digital audio tape (DAT), 145-147
heads, tape recorders, 56-61
hexadecimal numbering system, 79-81
hi-fi (*see* stereo recordings)
high-frequency roll off, 63

K

Kenwood I-Z9 home DAT recorder, 193
Kenwood KDT-99R automotive DAT deck, 182-185

L

lasers, compact discs (CD), 109, 113-118, 124-127
   coherent vs. incoherent light, 114

lasers, compact discs (CD) (con't)
  phase shift, 115-117
  phase, 114-117
  sine waves, 114
  white noise, 114
linearity
  digital audio tape (DAT), 99
  digital recordings, 87
liquid spills into DAT equipment, 228
loading problems, DAT equipment, 213-221
LPs (see disc recordings)

M

magnetic damage, 210
magnetic phonograph cartridges, 52-53
maintenance/troubleshooting
  cassette care, 205-210
  cassette loading, 208-209
  cleaning, 210-212
  condensation, 181-182, 209-210
  crunching noise, 225-228
  data display inoperative, 223
  dropped-equipment damage, 229
  ejection problems, 213-221
  labeling cassettes, 209-210
  liquid spills into equipment, 228
  loading problems, 213-221
  magnetic damage, 210
  noisy operation, 225-228
  program number search inoperative, 222
  search operations slow/erratic, 222-223
  smoke or sparks, 228
  sound distorted, 225-228
  sound out, 223-225
  sound problems, 223-228
  temperature extremes, 210
  track search inoperative, 222
manufacturers of DAT, 247-248
market analysis, future of DAT, 230-234
midrange speakers, 23
modulation techniques, 46-47, 83, 84
  amplitude modulation (AM), 85
  frequency shift keying (FSK), 84
  pulse amplitude modulation (PAM), 85
  pulse code modulation (PCM), 86-87, 98, 105-107, 135

pulse number modulation (PNM), 86
pulse position modulation (PPM), 86
pulse width modulation (PWM), 85-86
monaural/monophonic sound, 26
multiplexed signals, 98-99

N

Nakamichi home DAT recorder, 193-200
noisy operation, 225-228
notch filter copy protection, 171-176
numbering systems, digital recording basics, 76-80
nybbles, 79

O

octal numbering system, 79-80
overemphasis of sound, 24

P

PCM converters, 135
  VCRs and digital audio tape (DAT), 105-107
phase shift, 115-117
phase, 114-117
phonographs (see also disc recordings), 2, 49-56
  ceramic cartridges, 51-52
  crosstalk, 55
  cylinder-type, Edison, 3-7
  disc recordings, 9-11
  equalization, 55
  flutter, 55, 56
  grooves in records, 54-55
  magentic cartridges, 52-53
  speakers, 22-25
  speed (rpm) standardization, 12-16
  speed control, 7, 11-16
  wax-cylinder phonographs, 8
  wow, 55, 56
playback heads, tape recorders, 59-60
price comparison, digital audio tape (DAT), 233-234
pulse amplitude modulation (PAM), 47, 85
pulse code modulation (PCM), 46-47, 86-87, 98, 135
  converters, VCRs, 105-107
pulse number modulation (PNM), 86

pulse position modulation (PPM), 86
pulse waves, digital recording basics, 84-85
pulse width modulation (PWM), 47, 85-86

Q

quadrophonic systems, 41-44
quantization errors, 91-92
    digital audio tape (DAT), 137
    digital recordings, 107, 137

R

R-DAT, 139-141
record bias
    digital audio tape (DAT), 99
    tape recorders, 72-74
record heads, tape recorders, 56-59
recorder, DAT, 141-143, 188-200
    Casio DA-2, 189-192
    Kenwood I-Z9, 193
    Nakamichi 1000, 193-200
recording techniques, 1
    analog recording basics, 49-74
    audio cassettes, 38-41
    compact discs (CD), 2, 47-48, 109-134
    digital audio tape (DAT), 48, 135-165
    digital mastering, 112-113
    digital recording basics, 75-108
    direct-to-disc, 111-113
    disc recordings, 3, 9-17, 49-56
    Edison's cylinder phonograph, 3-7
    eight-track tape cartridges, 36-38
    equalization, 68-72
    modulation techniques, 46-47
    phonographs, 2, 49-56
    pulse code modulation (PCM), 46-47
    quadrophonic sound, 41-44
    reel-to-reel, 30-35
    signal-to-noise ratios, 71
    stereo recordings, 25-30
    tape recordings, 18-22, 30-41, 56-74
    tape-to-disc, 21-22
    video recordings, 44-46
    wax-cylinder phonographs, 8
    wire recorders, 17-18
records (see disc recordings)

reel-to-reel recordings, 30-35
    speed control, 64-66
    stereo, 33-35
reflection of sound, 24
resolution, digital recording basics, 90-92
roll-off, high-frequency, 63
rotating-head DAT recorders, 99-101, 103-105

S

S-DAT, 139-141
sampling frequency
    digital audio tape (DAT), 155-160
    digital recording basics, 87-90
SCMS copy protection, 176-178, 234-235
signal sampling, digital audio tape (DAT), 155-160
signal-to-noise ratio
    digital recording basics, 86, 91
    tape recordings, 71
sine waves, 114-115
smoke from DAT equipment, 228
sound problems, 223-228
SPARS code, compact discs (CD), 131-132
speakers, 22-25
    acoustics of rooms, 24
    crossover networks, 24
    distortion, 24
    frequency response, 25, 62
    midrange, 23
    quadrophonic sound, 41-44
    stereo systems, 25-30
    tweeters, 23
    voltage signals, 66-68
    woofers, 23
specifications, DAT, 141-43, 160-165, 200-203
stationary-head DAT recorders, 99, 101-103
stereo recordings, 25-30
    audio cassettes, 40
    digital audio tape (DAT), 153-154
    disc recordings, 54-55
    quadrophonic sound, 41-44
    reel-to-reel, 33-35

T

tape recordings, 18-22, 30-41, 56-74, 83

tape recordings (con't)
  audio cassettes, 38-41
  editing techniques, 19-21
  eight-track cartridges, 36-38
  equalization, 68-72
  erase heads, 60
  flutter, 31, 74
  flux, 67-68
  frequency response, 61-66
  head arrangements, 60-61
  playback heads, 59-60
  record bias, 72-74
  record heads, 56-59
  reel-to-reel, 30-35
  signal-to-noise ratio, 71
  speed control, 64-66
  tape-to-disc recording, 21-22
  voltage signals, 66-68
  wow, 31, 74
tape requirements, digital audio tape (DAT), 143-145
temperature control, 210
total harmonic distortion (THD), 128, 201-202
troubleshooting (see maintenance/troubleshooting)
tweeters, 23

V

VCRs, 44-46
  formats and standards, 136
  PCM converters, digital audio tape (DAT), 105-107, 135
vibration and sound waves, 1-2
video recordings, 44-46
  PCM converters, 135
  VCR format standards, 136
  videodiscs, compact discs (CD), 118-119
videodiscs, 118-119
voltage signals, tape recordings, 66-68

W

water damage, 228
wax-cylinder phonographs, 8
white noise, 114
wire recorders, 17-18
woofers, 23
wow
  compact discs (CD), 128
  digital audio tape (DAT), 201
  disc recordings, 55, 56
  tape recordings, 31, 74